# THE HOME DISTILLER'S GUIDE TO SPIRITS

### REVIVING THE ART OF HOME DISTILLING

#### Steve Coomes

# THE HOME DISTILLER'S GUIDE TO SPIRITS

## REVIVING THE ART OF HOME DISTILLING

### Steve Coomes

FIREFLY BOOKS

# A FIREFLY BOOK

Published by Firefly Books Ltd, 2015

Copyright © 2015 Quintet Publishing

All rights reserved. No part of this publication may be reproduced, stored in a retrieval system, or transmitted in any form or by any means, electronic, mechanical, photocopying, recording or otherwise, without the prior written permission of the Publisher.

First printing

**Publisher Cataloging-in-Publication Data (U.S.)**

Coomes, Steve
    The home distiller's guide to spirits/Steve Coomes
160 pages : color photographs; cm
Includes index.
Summary: The beginner's guide to making your own moonshine, fermented drinks, and more. With more than
50 original photography and direction on distilling, bottling and canning.
ISBN-13: 978-1-77085-477-2
1. Distilleries – Handbooks, manuals etc.  2. Distillation apparatus – Handbooks, manuals etc. I. Title.
663.16 dc23    TP597.C6654    2015

**Library and Archives Canada Cataloguing in Publication**

Coomes, Steve
    The home distiller's guide to spirits/Steve Coomes
Includes index.
ISBN-13: 978-1-77085-477-2 (bound)
1. Distilleries – Handbooks, manuals etc.  2. Distillation apparatus – Handbooks, manuals etc. I. Title. TP597.C65 2015
663'.16    C2015-900736-4

Published in the United States by
Firefly Books (U.S.) Inc.
P.O. Box 1338, Ellicott Station
Buffalo, New York 14205

Published in Canada by
Firefly Books Ltd.
50 Staples Avenue, Unit 1
Richmond Hill, Ontario L4B 0A7

Cover design: Erin R. Holmes
Printed in China by RR Donnelley

Conceived, designed, and produced by Quintet Publishing:
Project Editor: Caroline Elliker
Designer: Rod Teasdale
Art Director: Michael Charles
Editorial Director: Alana Smythe
Publisher: Mark Searle

## Image credits

l = left, r = right, t = top, m = middle, b = bottom

**Alamy**: Tim E. White 45; ZUMA Press, Inc. 47
**Big O Ginger Liqueur/OTT Enterprises LLC.**: Angela Castelli 146; Brandon Voges of Bruton Stroube Studios 147
**Boord & Son Distillery**: 101r
**Boundary Oak Distillery**: 58–9
**Brewhaus America Inc.**: Mary Polakovic 24–5
**Copper & Kings American Brandy Company**: 125–9
**Corbis**: Molly Riley/Reuters 7; Jonathan Blair 35; Everett Kennedy Brown 38; Regis Duvignau 40t; Peter Hudec 40b
**Corsair Artisan, LLC.**: Andrea Behrends 106–7
**Distilled Spirits Epicenter**: 46
**Ian Garlick**: 55, 56, 77–8, 87, 92, 97, 104–5, 109–10
**Getty Images**: William Hogarth 8; Buyenlarge 11; ferrantraite 23l; John Burke 31; Dorling Kindersley 32; National Archives (digital version Science Faction) 33; Zero Creatives 34; Dewald Kirsten 36; catnap72 37; Juan Silva 81; Jamie Grill 91; Bill Boch 95, 139; Stefano Scata 121r; Nicolas Thibaut 122; AFP 123; Judd Pilossof 125; Jonathan Irish 141; Tafari K. Stevenson-Howard 142; Jeff Kauck 143l; Chicago Tribune 144; Simon Watson 145; Cultura Travel/Philip Lee Harvey 151
**Hauraki Brewery Company Limited**: 50–1
**Haymans Gin**: Anthony Cullen 101l
**Hillbilly Stills Distilling Equipment & Supplies**: Socially Present 19; Mixer Direct 28–9
**Library of Congress**: Keppler & Schwarzmann 13
**Merchant House/The London Bar Consultants**: Angela Elliker 114–5
**Ole Smoky Tennessee Moonshine**: 17, 152–3
**Science Photo Library**: Veronique Leplat 23r
Steve Coomes: 83l, 83r, 84, 116
**StockFood Ltd.**: Martina Schindler 143r; Mikkel Adsbol 148; Rockard Forsberg 149
**Still Waters Distillery**: 72–3
**Syntax Spirits**: Jafe Parsons 85b, 89; Jeff Copeland 88
**Thomson Whisky New Zealand**: Max Thomson 16
**Jon Whitaker**: 49, 52, 60, 74, 113, 131–3, 135–6

While every effort has been made to credit contributors, Quintet Publishing would like to apologise should there have been any omissions or errors, and would be pleased to make the appropriate correction for future editions of the book.

The author and publishers of this book do not condone the illegal production of distilled spirits. It is the responsibility of the reader to determine whether the distillation of spirits is legal in the jurisdiction in which they propose to distill products, to obtain the necessary licences and to refrain from engaging in illegal distillation. The author and publishers of this book take no responsibility for any damage to either persons or property which may occur as a result of practicing unsafe distillation, or for any liability, criminal or civil, incurred by any reader as a result of engaging in distillation or the distribution of distilled products.

# CONTENTS

INTRODUCTION 6

EQUIPMENT 18

TECHNIQUES 30

VODKA 42

WHISKEY 62

RUM 80

GIN 98

BRANDY 118

INFUSIONS 138

GLOSSARY 154

DISTILLING RESOURCES 157

INDEX 158

ACKNOWLEDGMENTS 160

# INTRODUCTION

The distillation of spirits has been practiced at home for centuries, and although laws and markets have changed over time, the basic principles of distilling haven't. In this book you will find a fascinating history of the process, advice on everything you need to know to guide you on your journey into the exciting world of spirit production, and tempting recipes to enable you to truly enjoy the fruits of your labor. The legal advice on the pages that immediately follow will ensure you remain on the right side of the law the whole time.

**RIGHT:** *An actor takes the part of George Washington in his distillery on the Mount Vernon plantation in Virginia.*

*Civilization begins with distillation.*

— WILLIAM FAULKNER

# DISTILLING AT HOME

Little more than 400 years ago, small-batch production of liquor was a way of life and the making of potable spirits (as well as beer and wine) an everyday chore on the 16th century farm. Excess grain and fruit was fermented and distilled into shelf stable beverages that could be traded for other goods, or just enjoyed.

"Distillation became an important part of many cultures because what they made had so many uses," says Michael Veach, a bourbon historian in Louisville, Kentucky. "Whether they made brandy from fruit or whiskey from grain, people learned it was much safer to drink fresh water with one of those in it rather than drinking that water straight. It was good for medicine, barter and general ease of life, so it was readily made by a lot of people."

Production increased dramatically as distilling became commercialized, and soon brandy spilled off French stills, whiskey bubbled over on every British Isle, vodka poured in from Eastern Europe, and genever (the original gin) was being shipped from Holland to nearly all of Europe.

In the New World, British colonists drank lots of rum imported from the Caribbean throughout the 1600s, yet it didn't take long before they started making it themselves from sourced molasses and cane sugar. By 1657, Boston, Massachusetts, had a rum distillery, and by the early 1700s, there were commercial rum distilleries from New England to the Carolinas. Despite the increasing retail accessibility of spirits, distilling at home continued to thrive, both in the United States and in Europe.

The 1700s saw the English master gin production, and it was made there in such extraordinary quantities that its overconsumption threatened the health of thousands. According to Lesley Jacobs Solmonson in *Gin: A Global History* (Reaktion, 2102), the tippling was simply out of control. "Never before had England been so mesmerized by a beverage; never again would the city of London be as consistently intoxicated as it was from 1720 to 1751. Indeed, from 1684 to 1710, while beer production fell by 12 percent and strong beer by 2.5 percent, gin production rose by 400 percent." And those are just the legally reported numbers.

Changes to England's distilling laws in 1690 broke the monopoly of large distillers on gin production, flooding the market with new micro-distillers and heaps of inexpensive gin that the poor guzzled with

**ABOVE:** *William Hogarth's* Gin Lane, *1751, an etching and engraving depicting the social evils of gin consumption.*

**ABOVE:** *The rebels escorting a tarred and feathered tax collector during the Whisky Rebellion, western Pennsylvania, 1794.*

abandon. Solmonson writes, "Some scholars have likened the Gin Craze, as it has become known, to the crack epidemic in 1980s America." The problem worsened as greedy distillers stretched their liquor with liquids not meant for human consumption, like methanol and sulfuric acid. Such bad liquor earned the name of "rotgut," appropriate since that's precisely what it did to those who drank it.

Faced with a public health disaster and nearly empty war coffers, Parliament sought to fix both problems simultaneously by regulating gin production through taxation and licensing. Though King Charles II created England's first liquor tax in 1660, it seemed to have had little effect on preventing bad production by small distillers. So, from 1729 to 1751, Parliament passed eight Gin Acts that further taxed the spirit's production and forced distillers to purchase licenses to produce spirits. Home distillers faced huge fines if caught, and informants were rewarded for tipping off the authorities.

In Colonial America at around the same time, a fascination with rum was soon to be replaced by a growing interest in whiskey made from grains grown in the eastern part of the land. Rye whiskey became exceedingly popular, followed by corn-based whiskey that was spiced with rye or softened with wheat. It didn't take long for American whiskey to become popular elsewhere; a great source of export revenue.

The popularity of whiskey drew attention to its taxability. In 1791, eight years after America won its freedom from England, the country still needed money to pay off its war debt, so Alexander Hamilton,

**ABOVE:** *Basic alcohol distillation equipment as it would have looked in Colonial America.*

Secretary of the Treasury, established the first tax on American distilled spirits: not only would distillers have to pay taxes on every gallon they distilled, but also a tax on the production capacity of their stills. The distillers did not accept the new tax willingly.

Washington, who surely thought his days of leading armies was behind him, found himself commanding 15,000 men to put down a large rebellion against the Whiskey Tax in effect in Carlisle, Pennsylvania. Washington's victory, however, only made America's illegal distillers that much more determined to break the new law.

As David W. Maurer wrote in his book, *Kentucky Moonshine* (University Press of Kentucky, 2009), "Nothing so stimulates the enterprise of human beings as to prohibit or penalize some activity," and indeed many thumbed their noses at the new regulations. They saw making tax-free liquor as a right of the freedom they'd fought for in the Revolutionary War, so they moved to distill their liquor out of sight and after daylight, which saw the activity dubbed "moonshining."

Clandestine distilling was nothing new. From the rugged Highlands of Scotland to the remote corners of America, distillers knew law enforcers would have to work hard to catch them, especially so in America, where moonshiners could easily move westward to avoid the law. Moonshining was particularly easy in Tennessee and Kentucky, where corn grew in abundance and limestone-filtered water gushed from springs and poured down creeks. Kentucky's extensive system of navigable waterways allowed for easy egress and ingress to the Ohio River, a broad stretch of liquid that merged easily with the Mississippi River just south of St. Louis, Missouri. Along that river flowed high-quality grain whiskey that worked its way down to its ultimate destination in New Orleans. There, some historians believe, it was named "bourbon."

By 1802, Thomas Jefferson's administration repealed the federal excise tax on spirits, although sadly that relief lasted little more than a decade. According to Maurer, in 1812 a tax of 9 cents was due on each gallon distilled and $2.70 per gallon of still capacity annually. And, for complying with the law, America's legal distillers were rewarded with tax increases. Just two years later, the tax jumped to 20 cents per gallon distilled, and by 1861, the first year of the country's bloody Civil War, the tax soared to $1.50 per gallon. By 1918, the tax rose to $2.30 per gallon before nearly tripling to $6.40 per gallon by the outset of the Prohibition in 1920.

Taxation wasn't the only force working against legal liquor makers. Temperance societies in England and the United States sought to curb excessive drinking by labeling high-proof spirits as the source of the problem in the early 1800s. Whether spearheaded by ladies who'd had enough of heavy drinking men, or growing religious fervor against alcohol consumption as a whole, such movements tilted popular opinion toward regulating alcohol production tightly and suggested eliminating its production altogether was a worthy goal.

**ABOVE:** *Onlookers watch as moonshiners use a large copper kettle still, funnels and jars to make and sell illegal liquor. United States, c. 1900.*

In the UK, the First World War saw the sale of alcohol curbed to dedicate resources for combat use: the Defence of the Realm Act passed in 1914 restricted pub hours and introduced a tax of a penny-per-pint. In Norway, alcohol was partially prohibited in 1917, with full-on prohibition kicking in by 1919, although thankfully it ended relatively quickly in 1926. Not so in Finland, however, where prohibition stretched 20 miserable years from 1915 to 1935.

In the U.S., after years of lobbying by The Anti-Saloon League and various other teetotaling interests, The Volstead Act was passed in 1919 to kick off the Prohibition in 1920. Legal manufacturing of beverage alcohol ended with the new law unless a distiller produced it for medicinal purposes. Over the next 13 years, until Prohibition was repealed, millions of prescriptions were liberally written for medicinal whiskey. Distillers such as Brown-Forman, creator of Old Forester Bourbon, might never have survived the Prohibition were it not for doctors' dubious support of liquor's curative powers.

In the grip of organized crime, illegal liquor making and smuggling flourished during the Prohibition, as the mob consolidated production and distribution. Drinking in secret gave rise to the speakeasy — establishments where people could enter for a tipple if they knew the right person or the right password. Police raids on such establishments were common, and on distilleries and breweries too, but new, better secreted versions, sprung up to take their place.

With Prohibition's end in 1933, the handful of distilleries that survived soon returned to formerly high levels of production. But they sold little whiskey at the outset because that which they made had to be aged. Moonshiners surely benefited from that temporary shortfall, though such small-time producers couldn't fill the gap either. Yet on they went, making and selling liquor in secret, or just making it for themselves.

Post-Prohibition, however, moonshiners weren't looked on as regular guys or friendly folks making liquor as a sideline. The activities of the mob and the massive government efforts to stop them sullied the reputation of anyone making illicit booze. Moonshiners were now viewed as criminals, and even if that line of work served to support their families, it wasn't dignified. Throughout the 1940s and 1950s, they were increasingly caricatured in comic strips and animated cartoons as hapless hill jacks who cared little for dignified work. But, of course, this new disdain didn't stop people from buying their products.

What surely slowed business for moonshiners was the proliferation of legal, safe and delicious manufactured liquor, products that were consistent every time one bought them. After all, why risk arrest buying moonshine that, even when made by an experienced distiller, won't taste as good as a well-made legal spirit costing roughly the same price?

There's no high art in making illegal 'shine. Moonshine has to be easy to produce it in secret — in less than ideal circumstances — and that usually means it's made from a "wash" of sugar and water. If ground grain is added, it's broadcast into the wash, but not "mashed" to extract its sugars through a chemical reaction brought on by cooking. The 'shine ultimately picks up some grain flavors and congeners, but no real character. Some call it corn whiskey, but it's not whiskey at all if the corn isn't mashed and fermented.

Mashing takes time and work. It's a mess to clean and sanitize fermenters and stills used for grain whiskey production, and when using crude stills, mash is easy to scorch. Most moonshiners who make booze to sell aren't in it for the craft; they're in it for money. So most often they resort to simple recipes based on sugar washes, the distillate from which is called "sugar shine." In truth, it's really a sugar vodka that can be made in about a week or less.

So, as Maurer asks in *Kentucky Moonshine*, "Why, when legal whiskey is available, do people make and drink moonshine?" If you're a customer, you drink it because it's unique, you like the unrefined taste, and you likely enjoy having something illicit — it's exciting to indulge in a little sin. If you're a business-minded moonshiner, you do it because it's cheap to make and can be sold at a good margin tax-free — the cost of ingredients is low and they're readily accessible.

Moonshiners like making hooch in the same way woodworkers like making furniture. It's a pleasurable task not to be hurried; you have to slow down to make it and doing it with friends can be fun. It takes time to turn sugar into alcohol. Then there's the modern moonshiner, the person who likes to make booze and consume it himself, or perhaps share it with friends.

Just as interest in wine in the 1980s and craft beer in the 1990s led many to make wine and beer at home, the spirits and cocktails boom of the new millennium has spurred interest in distilling spirits at home. That's making illicit liquor, though few, if anyone, outside the law would see as a criminal act.

As of the writing of this book, New Zealand is the only country in the world where hobby distilling is legal. The country has allowed it since 1996, and despite initial concerns that tax evasion, poisonous liquor production and stills explosions would result, all have proved unfounded. The only way to get in trouble distilling in New Zealand is to get caught selling it.

The good news is hobby distilling is beginning to find favor in the United States. In 2013, the Hobby Distillers Association (see page 24) was created to lobby legislators at every level to consider legalizing the act, and sources say lawmakers are receptive to their arguments thus far.

This book is dedicated to those distilling at home for fun, for the pure and simple pleasure of doing it, as well as to those who are contemplating doing so and are keen to learn more. To you all, we offer this text and raise (a legal) glass of spirits.

**RIGHT:** The Moonshiners of Capitol Hill, *cover of Puck, v. 65, no. 1665 (1909 January 27) by L.M. Glackens.*

# IS IT LEGAL?

The only country in the entire world that allows legal hobby distillation of potable spirits is New Zealand. (We envy you, Kiwis!) Whether that will change anytime soon no one knows, but there is hope.

The issue, of course, isn't whether one owns a still at home, it's what he or she does with it. Many countries allow the hoi polloi to operate stills as long as the activity is truly unexciting, such as purifying water or extracting essential plant oils.

In the United States you can even distill ethanol legally at home *as long as it's for fuel*. But if you make it potable, either micro-distilling or as a hobby, and are caught manufacturing or selling the finished product, you're likely facing prison time, stiff fines, or both. The U.S.'s first president, George Washington, was a talented home distiller who made a lot of money making rye whiskey, and even though he added the first excise tax on American-made whiskey, he at least allowed citizens to make it at home. But since the Prohibition of 1920, making hooch at home has been forbidden.

On the other hand, craft distilling, as a business, is legal in virtually all U.S. states. You can obtain a license and other required permissions to run a commercial distillery that sells bottled liquor for payment. High fees based on the traditional still size and production of a large commercial distillery have made it prohibitive for craft distillers. The rules are being changed in many places, however, to allow for smaller stills and lower production. Some states, e.g., Oregon, are more advanced in this than others. Zoning by-laws are also being changed to allow distilleries and some municipalities have clued in that a distillery that is allowed to sell samples, like a winery, can be good for tourism.

Hobby distillation of beverage spirits happens even in the most heavily regulated environments. Despite the risks involved, it's clear that people want to make

**In 2011**, TheTelegraph.com reported that five men were killed when an illegal vodka still exploded in Boston, in the English county of Lincolnshire. Though reports at the time did not detail the operation's size, descriptions of the "massive explosion" implied at its center was a very large still run by Eastern European immigrants working clandestinely. Quick police work also determined their vodka had been sold to neighbors and nearby liquor stores.

**Also in 2011**, the DailyMail.com reported on a resident of South Wales who was found dazed and walking outside his home in "blackened underpants" following a makeshift vodka still explosion — which he operated in his bedroom. His neighbor heard the explosion, saw him outside, noticed his badly burned arms and charred boxers, alerted paramedics and tended to his wounds.

**In 2008**, Alabama.com reported on three men in Athens, Ga., who sought treatment for burns that doctors found suspicious. Aware the sheriff was on the way, they admitted that their 25 gallon illegal still exploded and showered them with hot liquor mash.

**RIGHT:** *A small copper grappa still designed for hobby distillation at home.*

**ABOVE:** *Mathew Thomson, head distiller and co-founder of Thomson Whisky New Zealand, Ltd.*

their own liquor at home. Though no formal survey can be undertaken, it's accepted that most only seek to make a little liquor to drink at home and share with friends. Selling illegal booze is most often the pursuit of criminals, though undoubtedly amateurs with hobby stills capable of producing gallons of liquor per run find themselves compelled to make a little unreported income now and then. And that's why it's illegal almost everywhere: governments want a cut of the action, so they insist that any liquor made must be done so above board. Unreported income is unreported income, plain and simple.

Often governments argue that laws against hobby distilling are designed to protect hobbyists from making dangerous drink or being maimed operating poorly made stills (see page 14).

Yet even hard-core moonshiners say you really have to screw things up to make either the process or the product dangerous. What with numerous books on the subject, multiple home distilling forums, plus endless YouTube video tutorials, it's not as though information on safe distilling is scarce. "Smart folks know to seek expert advice," says Rick Morris, owner of Brewhaus, a distilling and brewing supply store in Keller, Texas (see page 24). Morris, who founded the Hobby Distillers Association, a lobbyist group working to change U.S. home distilling laws, adds: "No matter what you use it for, you're distilling fuel, which has its risks. I think part of what would help increase safety even more is if people felt free to talk about it openly."

Which brings us back to **New Zealand,** where Kiwis can make alcohol at home anytime. According to Morris, this move to legalize hobby distilling began in the 1980s, when its Labour Government sold off many departments to be run as private enterprises. Prior to that, the country's Customs Department rode hard on illegal distilling; no small task for an increasingly smaller government. Customs considered making citizens pay an excise tax to distill alcohol, but it also deemed that effort beyond its control. So in 1996, micro-distilling became legal there.

New Zealand's only demand on micro-distillers is that their spirits are made for personal consumption. According to Peter Wheeler, co-owner of Hauraki BCL, "The law says you can't even so much as hand that distillate across the fence to your neighbor, though no one monitors it that closely of course." About the only way to risk real trouble there, he adds, is to sell your liquor, "That's a no-no."

In nearby **Australia**, the laws are starkly different. Despite what experts say is an active hobby distilling culture there, personal distillation of spirits is not allowed. Australians can have stills, as long as they're no larger than 5 liters and used only for distillation of water or extraction of essential plant oils.

**Canada** is similarly restrictive. The Customs and Revenue Agency say that if Canadians want to distill anything other than water, they need a license. Even for water only, the still must meet strict specifications to avoid suspicion.

**ABOVE:** *Moonshine on sale at Ole Smoky Moonshine distillery, Gatlinbrg, Tennessee, founded in 2010. Master distiller Justin King and his team continue the family tradition of distilling for commercial purposes (see page 152).*

The situation is similar in the **United States**, home to a large, underground hobby distilling community. According to the U.S. Alcohol and Tobacco Tax and Trade Bureau's (TTB) website, Americans "may not produce spirits for beverage purposes without paying taxes and without prior approval of paperwork to operate a distilled spirits plant." In other words, go pro or go to jail. Americans can own a working still without a permit as long as it's used for distilling water, extracting essential plant oils, or producing ethanol for fuel. To follow the law, one must apply for permits to produce it.

But here's the catch: If someone buys a premade still, those manufacturers must submit that customer's personal information to the TTB. And TTB officers, if suspicious, can drop by unannounced and ask to see it. It's widely claimed such inspections happen infrequently. Not only are reports of home still busts far and few between, the TTB likely has larger problems to address. Still, the threat of inspection is always there, and penalties are stiff if one is caught.

The rules aren't any less liberal in the **United Kingdom**. Under the Alcoholic Liquor Duties Act of 1979, it is illegal to distill spirits for personal use. Precisely, "No person shall manufacture spirits, whether by distillation of a fermented liquor or by any other process, unless he holds an excise licence for that purpose under this section." Translation: You must have a license to distill because we want to tax your revenue. There is one exception: distillation without an excise license is allowed if it's for medical or scientific purposes, not for human consumption.

# EQUIPMENT

If you've not begun distilling at home — and we're not encouraging you to do so illegally — you might look at the distilling equipment list included in this chapter and think this hobby too pricey or detailed. Truth is it can be both, which is why many hobby distillers get their start making beer and wine. Both are far less costly and time intensive.

But you have to start somewhere — and the list provided, though by no means exhaustive, covers the basics for doing just that by describing most of the tools and parts required to distill liquor. All are widely available on the Internet and, when purchased over time, fairly affordable. (Like any hobby, distilling can become outrageously expensive if indulged without fiscal prudence.) And what's great is the fact that you needn't have the latest and greatest tools to make a fine product. As long as you have the essentials and the diligence to follow the advice of experts — not to mention the myriad recipes shared online — you can distill potable spirits as it's been done for centuries. See page 157 for a comprehensive list of worldwide distilling resources.

**RIGHT:** *Show room at Hillbilly Stills distillery, Barlow, Kentucky (see page 28).*

They liked me so long as the liquor flowed at my house, but I haven't seen any of them around lately.

– BUD ABBOTT, ACTOR

# CHOICE OF STILLS

Anyone who took chemistry in school ran distilling experiments in much the same way as the Egyptians did when they created the alembic still in 800 A.D. Liquid was heated to vapor in one vessel, which traveled through a tube to another vessel, where it was purified, concentrated and condensed.

Although there have been endless functional improvements since the invention of stills, they work much the same way today, only the results are more refined and precisely controlled. Distillers have polished their craft and tooled their hardware to a degree that the finished liquid they're after is nearly identical with each distillation run. Despite those advances, "it still takes someone to turn the dials," as they say in commercial distilleries, and that human input begins with the choice of stills for hobbyists.

## POT STILL

Heat is applied directly to the pot, which contains the wash or mash. Alcohol and water vapors then rise through its cap and to a lyne arm, which collects them and directs them toward a condenser. The condenser is often a copper coil called a "worm" that is immersed in a tub of cold water. The water's low temperature condenses the vapors into liquid called "low wines," a low-proof (about 50 proof) blend of ethanol, water and other alcohols. That liquid trickles from the open end of the worm and into a collection vessel to complete what is called the stripping run.

After emptying the pot of any residue from the stripping run, the low wines is returned to the pot and heated again to begin the finishing, or spirit, run. This time higher-proof spirits (about 130 proof) emerge from the still ready for blending, aging or bottling.

Pot still distillation is beloved and bemoaned for its imprecision. Cuts made on spirits produced in these stills are less exact than those made on column stills. But pot distillation produces spirits with full body, rich mouthfeel and well-rounded flavors that sometimes are "refined out" in column distillation. Without a doubt, pot distillation puts far more control of the spirit's outcome in the hands of the distiller.

For hobbyists, pot stills come in sizes small enough to fit atop a kitchen stove, and vessels able to run several hundred gallons of wash at once.

## COLUMN STILL

In the strictest sense, column stills for hobbyists are essentially hybrid pot stills with a reflux column attached to the top of the pot and then connected to a condensing apparatus.

Just as happens with pot stills, the mash or wash is heated directly in the pot, which causes alcohol and water vapors to rise. But this is where the column's influence becomes profound: placed inside the column at specific intervals is copper mesh packing or perforated copper plates that cause the vaporized distillate to reflux. That means as rising vapors hit those layers of cooler metal, most of the ethanol vapor moves onto the condenser, while water and other unwanted alcohols condense and fall back into the pot. Those liquids are again vaporized and refluxed multiple times to yield a pure, high-proof spirit (160 proof to 190 proof) in one distillation run.

Column stills allow the distiller a high degree of precision when making heads, hearts and tails cuts. The result is leaner distillate that bears bright and well-defined flavors. However, some argue fairly that such spirits lack the richer mouthfeel of pot distilled spirits.

For hobbyists, pot stills with columns are usually large enough to run anywhere from a gallon to several hundred gallons of wash at once, which will generally suffice for the desired quantities.

**ABOVE:** *Small distillery pot still: ancient and all-purpose, although lacking in precision.*

**ABOVE:** *Traditional double column still, a hybrid pot still that produces a leaner distillate and more distinct flavors.*

# DISTILLING EQUIPMENT

For information on where to source equipment, see Distilling Resources (page 157). For useful explanations of distillery terms, see Glossary (page 154).

**Airlock.** A one-way valve that, during fermentation, allows $CO_2$ gas to escape while blocking outside air from entering the fermenter.

**Alcoholmeter.** An instrument calibrated to read a liquid's density based on its percentage of alcohol. This is used to measure the alcohol content in a finished spirit.

**Boiler.** Device in which a wash or mash is heated for distillation. SEE ALSO KETTLE.

**Brew pot.** Large metal cooking pot in which washes and mashes are heated to encourage liquefaction and saccharification. SEE ALSO MASH POT.

**Cap.** The removable top piece or assembly of pieces on a traditional moonshine still. On a pot still, this is typically called the head.

**Cap arm.** Copper pipe connecting to the still cap for the purpose of conveying vapors to the condenser. SEE ALSO LYNE ARM.

**Collectors/collection vessels.** Containers used to collect condensed alcohol as it runs off the still.

**Column.** A metal (usually copper) pipe attached to the top of a pot still through which vapors pass as a mash or wash is heated. It is connected to a condensing apparatus. Within the column is copper mesh or perforated metal plates through which vaporized alcohol flows to the condenser. The mesh and plates assist in increasing the purity and proof of the distillate by condensing water vapor but not alcohol vapor.

**Condenser.** A two-walled, sealed pipe with one end connected to the still, and the other submerged in cold water to rapidly reduce the temperature of alcohol vapor to condense it. More modern, small-scale condensers feature a copper pipe jacketed by another pipe through which cold water circulates to condense alcohol vapor.

**Doubler.** SEE THUMPER/THUMP KEG.

**Extractor.** A metal column through which vapors rise and are moved to a condenser.

**Fermenter.** A container for mixing and holding ingredients during fermentation. For home distillers, these commonly are food grade plastic buckets purchased from wine or beer making supply stores, or a glass "carboy," which is a giant jug with a sealable mouth that can be fitted with an airlock valve.

**Graduated cylinder.** SEE TEST CYLINDER.

**Hydrometer.** An instrument for measuring the density of a liquid relative to pure water. In distilling, a hydrometer is used before fermentation to calculate the potential alcohol in a solution by measuring the water's density relative to the percentage of sucrose suspended within it. It is used again after fermentation to recalculate potential alcohol.

**Jacketed condenser.** A two-walled pipe that condenses alcohol vapor. Hot vapor moves through the inner pipe, while cold water circulates in the outer pipe and condenses the vapor.

**Kettle.** Device in which a wash or mash is heated for distillation. SEE ALSO BOILER.

**Lyne arm.** The metal tube or "arm" running from the head of a pot still to a condenser. SEE ALSO CAP ARM.

**Mash pot.** Large metal cooking pot in which washes and mashes are heated to effect liquefaction and saccharification. SEE ALSO BREW POT.

**Packing.** Material such as copper mesh or ceramic rings that are suspended in a column still to promote refluxing.

**Parrot.** A holding vessel located between a still's condenser and its collection container. It catches distillate for the purpose of measuring its alcohol

**ABOVE:** *Copper coil worm inside the still's condensing chamber, where vapor in the worm is turned into liquid alcohol.*

**ABOVE:** *Airlock valve allowing $CO_2$ to escape.*

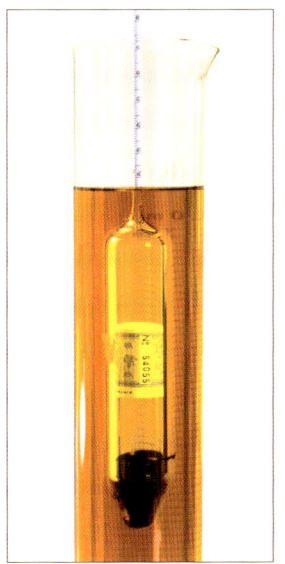

**ABOVE:** *Alcoholmeter in measuring alcohol content.*

content with an alcoholmeter. The parrot allows alcohol content readings to be made during the entire distillation run.

- **Pot still.** A still in which heat is applied directly to a metal pot containing wash or mash distilled in batches (rather than continuously by a column still). Historically known as the alembic still, pot stills are commonly used for cognac, Scotch and Irish whiskey and rum production.
- **Racking cane.** A siphon used to transfer fermented wash or mash to a still.
- **Raschig rings.** Tiny ceramic rings suspended in a still column to promote refluxing.
- **Reflux column.** A metal tube ascending from a still boiler wherein alcohol is refluxed to increase proof (the ethanol content).
- **Reflux plates.** Perforated copper plates placed within a still column to encourage refluxing.
- **Sight glass.** Small windows cut into a still column allowing distillers to view activity within column.
- **Still cap.** A metal cap affixed to the top of the boiler to support a reflux column or a condenser assembly.
- **Stirring spoon.** Long, nonreactive plastic or wooden spoon used to stir the cooking mashes or agitate the ferments.
- **Test cylinder.** A tall, narrow cylinder sometimes referred to as a "graduated cylinder," used to hold a sample of liquid for testing with a hydrometer or an alcoholmeter.
- **Thumper/thump keg.** A sizable chamber (sometimes made from a wood barrel, sometimes metal) linked by copper tubing between a distiller and a condenser. It is used to increase the proof of distillate emerging from the still and eliminate the need for a finishing run. It gets its name from the thumping sound made by vapors coursing through condensed liquid inside it. It may also be referred to as a "doubler."
- **Vapor cone.** A metal cone resting between the boiler and the still column or swan neck that directs vapor into the column.
- **Worm.** A copper tube typically 15–20 feet (4.5–6 m) long that is connected to the lyne arm on the still, carefully coiled and placed in a vessel holding cold water that condenses alcohol vapors passing through the coil.

# WORDS OF WISDOM

## RICK MORRIS
### OWNER, BREWHAUS
Keller, Texas
www.brewhaus.com

For 25 years, Rick Morris has made alcohol in the form of beer, wine and distilled spirits as a hobby. In 1992, he turned that hobby into a business called Brewhaus, which supplies home alcohol producers with everything they need. In 2013 Morris launched a significantly different and potentially more powerful venture, the Hobby Distillers Association (HDA). Its goal is to lobby American legislators to change post-Prohibition laws that would allow distillation of potable spirits at home. Morris concedes that the battle could be a long one, but favorable reception by multiple legislators, both in his home state of Texas and in Washington, D.C., has him believing hobby distillation will eventually become legal in the United States.

### What was the catalyst that led you to create the HDA?
In 2013, the TTB (Alcohol and Tobacco Tax and Trade Bureau) came to us and our competitors and required us to start preparing lists of customers who bought stills, kettles and columns from us. They wanted to have a better idea of who might be distilling spirits at home and that got people a little fired up.

When they made some busts in Florida in February of 2014, that's when a lot of people came on board and decided it was time to act now. That's when we formed the association, to have one concise and consistent voice that spoke clearly to our government and to create a collective effort that could be sustained long-term.

### How many members are there?
Roughly 1,300, and actually most of those signed up in the first few months. Our lobbyist said that response was huge, since most associations are lucky to get 200 to 300 members in their first year. And these people are putting names to our membership, which is a tremendous show of support.

### Any success to report so far?
We're making progress, but it's all happening at the speed of government. There are so many steps that you have to go through before you get in front of the right people.

🍶 **Do you believe that home distilling will be legal in the United States anytime soon?**

We're hoping so. We've been to Washington three times, and I have met with both of my state senators and representatives in the TTB. Soon we're going to meet with Treasury, and we have also met with Sen. John Cornyn, one of the chairs of the Senate Finance Committee.

🍶 **When you get in front of legislators, what do you tell them?**

Most of them don't even realize that hobby distilling is illegal. They know brewing and wine making at home are legal, and so they think distilling for personal use falls under the same category. They don't know that if you take wine and heat it up to make a brandy you're breaking the law. When we told that to Sen. Cornyn, he was surprised.

So far our message has been well received — we have yet to meet a senator or representative who's against it. But for any changes to be made, we have to get somebody in government to take the lead. Our lobbyist has gotten in there and met with the Senate Finance Committee and the House Ways and Means Committee, which is good.

🍶 **Is illegal distilling common?**

Judging by the amounts of yeast we're selling you can get an idea who's doing something they shouldn't be. Now I have no real proof, but people in the industry believe that it's just a very small amount of people who are distilling and selling it. That's part of why we think it should be legalized, because that would drive some of that away. If people can make it themselves, then they don't have to buy it illegally.

🍶 **Why do you think hobby distilling has grown so much lately?**

There's no doubt that the TV show *Moonshiners* has brought more attention to it. But what a lot of people don't know is that it's been very popular for quite some time. The show just brought more light to it.

**ABOVE:** *Completed distilling columns at Brewhaus, ready for shipping.*

🍶 **Give me a vague personality profile of the home distiller.**

It's not Tickle (a not-so-polished illegal distiller on *Moonshiners*). You've got doctors and lawyers — educated people — doing this. It's something fun to do, to fire up the still on Saturday and sit around and make liquor: it's not about cost savings for them — it's about the pride of putting a bottle of bourbon on the table and having friends say, "My, that's good!" That's the hobbyist, a person who wants to create something and share it with others.

# ESSENTIAL BAR TOOLS

It makes little sense to spend all your money on good distillation equipment only to not have good bar tools to finish the job.

**Bar spoon.** An indispensable and multi-use long-handled spoon created for stirring cocktails and retrieving fruit garnishes from jars. It holds about 1 teaspoon (5 ml) of liquid.

**Champagne flute.** An elegant long-stemmed glass with a narrow, tall bowl that's tapered at the opening.

**Collins glass.** Tall, clear cylindrical glass holding 10–12 ounces (300–350 ml), ideal for slow sipping.

**Coupe glass.** Short stemware whose 7½-ounce (220 ml) bowl resembles a flattened or sawn off wine glass; used for chilled cocktails that have the ice strained off.

**Cutting board.** Dedicate one for your bar so it won't be tainted by onion and garlic flavors left behind from food preparation; a small one will suffice for cutting up fruits and herbs. Wood boards are nice, but polypropylene is affordable and easy to clean.

**Highball glass.** A footed, 8–12 ounce (250–350 ml) glass designed to hold iced cocktails blended with carbonated beverages.

**Hurricane glass.** A tall, footed glass designed to hold 20 ounces (600 ml) or more of slow-sipping punch. The wide-mouthed glass is narrowed slightly at the center to make it visually elegant and easy to hold.

**Ice trays.** Great cocktails require hard and well-formed ice that cools liquid rapidly, but melts slowly. Modern flexible rubber trays are excellent tools to achieve this. Ice trays are made for sizes ranging from standard 1 inch (2.5 cm) square cubes and 4 inch (10 cm) long rectangles (for Collins glasses) to massive baseball-sized spheres. Shop around and find what suits you, but measure your glassware first to be sure the ice will fit.

**Jigger.** A small metal measuring container with two measurement cups (½–¾ ounce (5–7.5 ml) and 1–2 ounces (30–60 ml)) on either end. I prefer the taller, narrower Japanese-style jiggers to their wider-mouth counterparts because they're neater and cleaner to work with.

**Juice press, handheld.** A small press used to squeeze small fruits, such as limes and lemons, sometimes to order.

**Juicer, electric.** Used for squeezing larger fruits such as oranges and grapefruits, or larger volumes of lemons and limes. Manually operated professional juice presses are nice, but pricey and unnecessary. If you must have one, spend the money to get a good one. You'll be frustrated by cheap models.

**Julep cup.** A metal 12 ounce (350 ml) cup created to keep iced cocktails exceptionally chilled. These aren't inexpensive, but they are unrivaled for keeping cocktails cold.

**Knife, channel.** A unique "bladeless knife" whose cutting channel is designed to carve lemon and lime twists. It's not mandatory, but it is an elegant and affordable tool.

**Knife, paring.** An essential tool for cutting fruit for garnishes or squeezing. A utility knife 4–6 inch (10–15 cm) long works well, too.

**Lewis bag.** A canvas bag into which large pieces of ice are placed for cracking with a mallet. Where plastic bags tear quickly, this bag won't.

**Mallet.** A blunt and broad-faced wood or metal hammer used for cracking ice.

**Martini glass.** SEE UP GLASS.

**Mixing glass.** A large, wide-mouthed glass used for mixing and stirring drinks to chill before pouring into the glass presented to the drinker. This is a completely optional tool, but a classy one for purists.

**Mug, heavy walled.** Earthenware or porcelain mug designed to retain heat of beverages, such as hot toddies and coffee drinks.

**Mug, Irish coffee.** A footed 8–10 ounce (250–300 ml) glass equipped with a handle for easy drinking of hot, and usually alcoholic, coffee drinks. Clear glass is used instead of opaque materials to boost each drink's visual appeal.

**Mug, Moscow Mule.** A squatty copper mug with a roomy handle designed to keep the Moscow Mule Siberia-cold.

**Muddler.** A small plastic, marble or wooden pestle used to mash fruit, fruit peels and herbs. In most cases, muddling is more about bruising than mashing; mashing can release bitter oils from herbs and citrus peels, so be delicate but firm.

**Peeler.** A vegetable peeler, preferably one that can be used with some precision and control.

**Rocks glass.** A squatty, wide-mouthed 10–12 ounce (300–350 ml) glass designed to hold cocktails served on ice.

**Shaker, cobbler.** A three-piece metal device consisting of a bottom tin that holds the cocktail mix, the top piece that covers the bottom tin but is equipped with a coarse ice strainer, and the cap, which holds the mixture in while you shake it.

**Shaker, weighted tin.** A single metal cup used for mixing a cocktail before shaking. Drinks are shaken using this bottom-weighted cup by inserting a smaller cup into its mouth to contain the cocktail while shaking.

**Shot glass, graduated.** Glass or plastic shot glasses for measuring. When compared to metal jiggers, these have several drawbacks: They don't pour as cleanly, they're slippery and they can chip and break. So let the souvenir shops sell these, and buy safer, more accurate metal jiggers.

**Speed pourers.** Pouring devices that insert into a spirit bottle's mouth and allow for rapid, neat and measured pours. If you like to work cleanly, these are indispensable tools.

**Strainer, Coco.** A fine sieve strainer designed to remove all particles from squeezed juices and herbs from drinks.

**Strainer, Hawthorne.** A two-piece strainer designed to allow liquid to pass through its flexible spring edge quickly while capturing ice or large particles, such as muddled fruits or herbs.

**Strainer, Julep.** A single-piece metal strainer with multiple uniform holes designed to let liquids pass through while retaining most fruit, herb or ice particles in the shaker cup.

**ABOVE:** *From left to right — bar spoon, Hawthorne strainer, channel knife, jigger, bottle opener, muddler and cobbler cocktail shaker.*

**Up glass.** A long- or short-stemmed glass with a wide-open triangular bowl used commonly for any drink served up (chilled but without ice). Though elegant and even sexy, the long-stemmed versions of these glasses tip over easily, so consider shorter-stem versions.

**Wine key/wine opener.** A collapsible three-tooled device used for opening bottles of wines. Good wine keys will have a small knife to cut through the foil or plastic liner at the bottle's top to access the cork. They'll also have a corkscrew to auger into the cork, and a lever that grips the bottle top's edge to assist in pulling the cork out cleanly. If you own a bi-levered corkscrew, pitch it in the trash and get a three-tooled key. If you don't know how to use a three-tool key, have an experienced server at a respectable restaurant show you.

# WORDS OF WISDOM

## MIKE HANEY
### PRESIDENT, HILLBILLY STILLS
Barlow, Kentucky
www.hillbillystills.com

Mike Haney was a hobby distiller who became determined to build a better still. After posting a YouTube video about his still, customers started calling — lots of them. The longtime paper-mill employee recognized the market demand and formed Hillbilly Stills in 2009. Revenue and production have skyrocketed ever since, and the firm's eight employees can't keep up with demand.

### How did you get into hobby distilling?

I wanted to learn the craft, so I bought a couple of stills from eBay, but they just didn't do what I wanted. I started studying commercial distilling equipment to see what made them so much more efficient. I took a good look at their reflux columns, all their different designs, in order to understand how those stills could get the job done in a single run. I kept asking myself, "Why can't this be scaled down to hobby level?" I eventually made a still with a 4 inch (10 cm) diameter column on top of a beer keg boiler. I even condensed it in my wife's above-ground pool, which is real redneck.

Pretty quickly I got it to where it was making 1 gallon (4 l) an hour. So I shot a video and put it on YouTube. I never had any intentions of starting a business or building stills, but people started calling.

### What led you to turn your hobby into a business?

My wife and I used to travel a lot, but when she was diagnosed with severe rheumatoid arthritis, that pretty much ended. All of a sudden, I had all this free time and needed something to do at home. One thing led to another, and she and I were sitting on the couch deciding whether I should quit my job so I could be at home with her more and do what really interested me.

### How have so many people found out about Hillbilly Stills?

We made a few videos along the way on how to make liquor at home, and we put up a website, though I didn't know a thing about search engine optimization. I made videos on how to make a simple mash, how to make cuts, how to run a pot still, how to convert a beer keg into a still, and all that got a lot of attention. The more I posted on YouTube, the more attention we got. We couldn't believe how many people were interested.

### Didn't the TTB (Alcohol and Tobacco Tax and Trade Bureau) frown on that?

The only reason that I could do those videos and put my face out there was because I had an ethanol producer's permit through the TTB. It was 100 percent legal and 100 percent free.

### Any idea of how many people are using your stills for beverage spirits production?

Oh, I would say 80 percent, and those are hobbyists. But as our business has grown, we've begun doing a lot more commercial work for serious craft distilleries.

### What is the smallest still that Hillbilly Stills makes?

Eight gallons (30 l), which is pretty small. What people don't always understand is that when you have an 8 gallon still, the most you're going to put into it is 6 or 7 gallons (23–26 l) of mash, and out of that you'll only get a half gallon (2 l) of hearts. That's a lot of work to make just a little bit. I think that a 13 gallon (50 l) still is the perfect size for an enthusiast. You can run that and make about three-quarters of a gallon (3 l) of hearts, and in a 26 gallon (100 l) still, you'll get about 1½ gallons (5 l). Choosing the size comes down to a question of how much you want to make for the time you invest.

### You sell complete still kits for the nonmechanically inclined. Are you highly mechanically inclined then?

Somewhat, but I sought a lot of help when it came to making stills. I credit Rob Sherman (at Vendome Copper & Brassworks, a large still manufacturer in Louisville, Kentucky) for a lot of what I've learned.

**ABOVE:** *Copper stills in the building process, Hillbilly Stills shop.*

I told him that I was starting this business on a very small scale and asked if he would help me. His exact words were, "Hell, yeah, I'll help you." That's not only another still builder in the United States: that's probably the best still builder in the United States.

### What do you believe has made the home distilling market take off worldwide?

There's a craft to it and it's an amazingly fun hobby. You can get together with buddies, try your own recipes and have a good afternoon doing it. No doubt that the allure of moonshine itself has really taken off and helped this along. Everywhere you look, moonshine is out there, on TV and the Internet. People can learn anything they need to know about it.

### Do you think it will become legal here in the United States?

I do, and I think it will happen in the next couple of years. Look at what homebrewing has done for the craft beer industry. I think the same would happen for the spirits industry with the boom in craft distilleries. There's a huge market to be tapped if the government would play ball. It may never happen of course, but if it did, it would be a big plus.

# TECHNIQUES

The irony of alcohol distillation is that it should be performed starkly sober, as H.L. Mencken states. Smart and safe distillation implies sober distillation, that allows for accurate monitoring of the numerous nuances of the craft. As many distillers will admit, boiling a wash is about as exciting as watching grass grow. So a distiller under the influence of his own spirits could find himself asleep at the still and messing up a run; or worse, doing a great deal of damage to himself and his equipment.

Our parents warned us to practice "Safety First," and our teachers and coaches taught us that doing anything well "is all about great technique." That's successful distillation in a nutshell. What follows is a brief overview of both points.

**RIGHT:** *Testing alcohol levels at a whiskey distillery.*

*The harsh, useful things of the world, from pulling teeth to digging potatoes, are best done by men who are as starkly sober... but the lovely and useless things, the charming and exhilarating things, are best done by men with, as the phrase is, a few sheets in the wind.*

— H.L. MENCKEN, PREJUDICES, FOURTH SERIES, 1924

# SAFE DISTILLATION

Ask any expert and they will tell you that distilling alcohol, when done smartly, is very safe. For the hobbyist, there are no mechanical processes involved in distilling, virtually nothing that can break when heating a mash or wash, in the boiling, collecting and condensing of those delicious vapors from the liquid.

But accidents do happen. Stills can leak liquid and alcohol vapor, distillers can be burned by the liquid or, in the worst of all cases, alcohol vapors leaking from the still can ignite, producing a virtually invisible flame that can be difficult to extinguish.

Again, when you ask experts why these accidents happen, more often than not they will say they are due to negligence, such as not preparing equipment correctly beforehand, and not monitoring the distillation process carefully.

The author and publishers of this book do not condone the illegal production of distilled spirits. We advise procuring the necessary licenses and refraining from engaging in unsafe practices. Since we know that hobby distilling takes place all over the world, we take a position of reporting on it and consulting the experts who do this legally. This advice comes from those who not only get paid to distill spirits for a living, but who do so safely. It is our hope that you will follow their example.

"When you think about the number of centuries that people distilled spirits over an open flame, you know that the process is safe if you're careful," says Tripp Stimson, Director of Distillery Operations, Kentucky Artisan Distillery in Crestwood, and a craft-distilling consultant. "Even real moonshiners who run into problems rarely talk about the bad things that can happen when you're not taking care of things: missing fingers, burns and still explosions. But if you just do this properly, you should be just fine."

For beginners, several sources strongly recommend the purchase of a premade still from a manufacturer with a solid reputation. While there is no shortage of YouTube videos demonstrating how to make stove top stills from cooking equipment or chemistry lab gear, Stimson, like others, points out that these produce such small amounts of alcohol that they aren't always worth using, given the time invested. "Like I keep saying, I'm not promoting the distillation of illegal spirits, but if you're going to put the work into it — and it takes some work — I think you should get a little bit better yield for your effort than just a pint or two." He adds, "If you

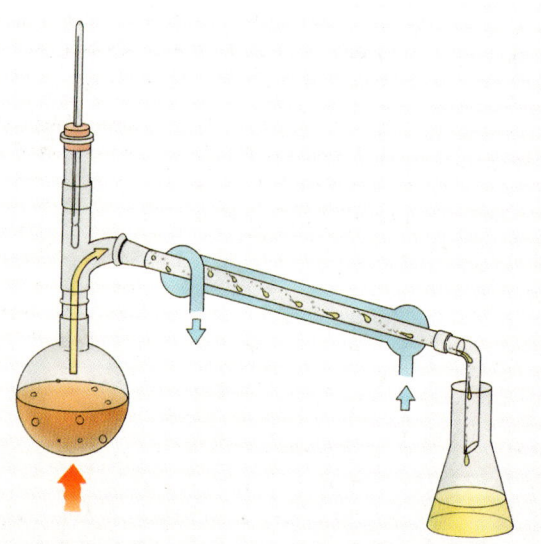

**ABOVE:** *The separation of mixtures using the distillation process; a simple process when managed correctly.*

do go the route of buying a still from a manufacturer, you'll probably get a decent size to work with to make it worth your while, and also get a piece of equipment made to run safely."

## ASSEMBLY

When you assemble the still, the safest way to test it is to add water to the boiler and run it as though you are distilling alcohol. If there are any leaks, the visible water vapor will warn you that the still is in need of tweaking before distilling any alcohol.

The chance that a manufactured still would be in such condition is low, says Mike Haney, president of Hillbilly Stills, in Barlow, Kentucky (see page 28), "A good still manufacturer will test all their products under air pressure even though a properly running still does not build pressure." He continues, "Sure, there could be a problem and that would be human error. But if it's properly tested, and we test our systems to 40 psi, it shouldn't happen. You're just moving vapor up a column, so there should be no pressure buildup at all."

For do-it-yourselfers, Haney says you must use lead-free solder (leaded solders are poisonous). "Five percent silver solder is the best, but that takes an oxyacetylene torch to get it right," he says. "Plumber's solder is fine, and you can work that with a propane torch. Just make sure that your solder is lead-free."

Haney says that for centuries moonshiners sealed leaks with a paste of flour and water; proving the near absence of pressure inside a properly running still and how low-tech distilling is. "And you can do that, too, but the point is to get the leak stopped," he says. "If you have a leak, the safest way to address it is to turn your heat source off and let the still cool completely so you can fix it the right way, by soldering it."

## HEAT SOURCES

Heat sources for stills vary mostly between propane gas burners, electric hot plates and electrical heating elements that are submerged in the liquid being distilled. Open flames must be used in well-ventilated areas. Both Stimson and Haney suggest that happens outside where possible; or at least in a garage or barn, with high ceilings and large doors that can be opened.

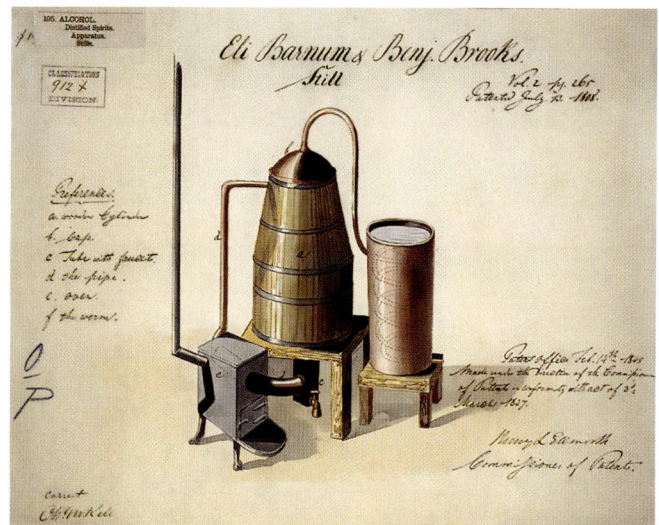

**ABOVE:** *A watercolor for a patent of a distillery designed by Eli Barnum and Benjamin Brooks.*

Haney says he is also sold on electrical heating devices for their precisely controlled heat output. Even a powerful propane burner is susceptible to outdoor breezes that can make heating tough to regulate. And an internal electrical element, he says, doesn't have that problem. "I hear people talk all the time about using electric (devices) to heat their stills in basements and garages," says Haney, adding that his company also sells propane burners for heating. "Using an electric element in the boiler almost idiot-proofs the whole process. Know what I'm saying? With gas, I'd want it vented really well. If it's not, it can kill you."

Stimson, Haney and many others interviewed for this book stressed that leaving a running still unattended is asking for trouble. As Stimson says, distilling can be boring or relaxing depending on your perspective, but in either case, a distiller must remain mindful of what's happening at all times to be truly safe: "The truth is it can be like watching paint dry because almost nothing is happening for a while, but then the action starts happening all at once. If I'm a beginner, I'm never leaving it. If you're experienced, you have to be comfortable in knowing what stage you're at and what's going on. But if you do leave it, only leave it briefly.

And it's never something that you can completely walk away from."

Both Stimson and Haney stress that although it may be boring waiting for distillate to run, these points in the process teach new distillers great lessons. Haney advises distillers learn to judge how hot the still becomes by briefly touching its exterior. At about 120°F (48°C) the exterior becomes hot. Not long after, he says, it will be too hot to touch; then you are near the point that the liquor is about to run.

**ABOVE:** *Safe distillation practiced in a controlled environment, with the right equipment.*

"You don't want to go anywhere for sure," Haney says. "You don't want a bunch of other activities going on. Relax, have fun, but pay attention."

The first distillate to emerge from the still will be methanol, since it has the lowest boiling point of all the liquids in the wash. Bottom line is: methanol is poisonous. Haney advises his customers to discard the first 2 percent of liquid that comes off the still. Stimson adds that, "Any time you're doing a small run, deciphering those heads cut will be extremely difficult. So definitely play it safe and get rid of that methanol by throwing out what comes off first."

Both men say that the best tool to detect when methanol is running is the nose. Expect not so lovely aromas, e.g., paint solvent and nail polish remover. "So when what's coming off starts to smell good, you're probably into your hearts cut," Stimson says, adding that the alcohol in this portion will also taste good. "Over time, as you gain experience, you will get very good at defining your cuts by taste and smell."

Haney allows that some distillers use alcoholmeters and thermometers to make their cuts, but he agrees that smell and taste are the best ways to do it. "Learn to feel it by rubbing it between your fingers," he says. "That's especially helpful in knowing when you are in your tails cut because it begins to get little bit oily and stinks. At every point, rub a little on your arm and smell it, which will tell you a lot. To me, though, tails are easy to smell. They smell like wet cardboard."

If steam rather than liquid is coming out of your condenser, you have one of two problems. The first is that your still is so hot that all the liquid inside is boiling, making it impossible for you to make cuts. An equally likely problem is that the water you're using to cool your condensing arm or coil is too warm to condense those water vapors. Whether you are using an old-fashioned copper coil condenser, called a worm, or a more modern space-saving heat exchange condenser (such as a shotgun condenser), a constant supply of cold water is essential to finish the process. "It's real simple: if you're puffing out vapor, the condensing water is not cold enough," Haney says.

Finally, to collect distillate safely, the collection vessel should be as far away as possible from the still itself. While a properly sealed still poses little risk to flammable, condensed alcohol, it's best to be safe by keeping it away from the heat source. Experts strongly advise the use of glass or stainless steel collection vessels only. Though it is potable, high-proof alcohol can be corrosive to some plastics, and Haney says plastics in general can lend an off flavor to distillate:

"Some are okay using plastic if it's cut below 100 proof. But for me, personally, I'm highly against that because I don't like any strange flavors coming off plastic. You can buy glass jugs pretty cheap at most stores, so why not use them?"

# FERMENTATION

Andreas Libavias, a 16th century German alchemist, described fermentation as "the exhalation of a substance through the admixture of a ferment which, by virtue of its spirit, penetrates the mass and transforms it into its own nature." If this makes your head hurt, then you're normal. He brilliantly reduced this enormously technical process to one sentence that, not surprisingly, confounds most nonscientists.

Fermentation is no simple matter, which is likely why distillery tour guides often stop their groups beside massive fermenters and say, "This where the magic happens." The good news is fermentation is actually simple to perform even if you don't know all the science. But we'll dig into it a little anyway.

Aside from aging, fermentation is almost the only function of spirits production that happens largely beyond the influence of distillers. They can start it and control it, but that chemical transformation Libavias wrote about is responsible for most of a distillate's flavor. The mere combination of water, sugar and yeast seems rudimentary, but it's so much more.

## YEAST: THE KEY TO FERMENTATION

As most of us learned in high school chemistry class, or perhaps later at the doctor's office, yeast is a fungus, a single-celled organism manifested in myriad forms. Its role in fermentation is clearly defined: to be added to a water solution, seek out sugars and turn them into alcohol.

The most common yeasts used in distilling are *saccharomyces cerevisiae* and *saccharomyces bayanus*, essentially two umbrella names for several other strains

**BELOW:** *A vat of fermenting rum at the Mount Gay Rum Refinery in Barbados.*

of yeast. When you visit brewing or winemaking supply outlets and look at the varieties available, you'll see what I mean. There are whiskey yeasts, rum yeasts, baker's yeasts, dry active distiller's yeast and more. There's even the turbo yeast, an all-in-one yeast and nutrient combination that conducts a rapid and complete fermentation. (Though I won't get into the technical discussion of it, it's important to know that distillers are divided on its positive or negative effects on the flavor of a spirit. What everyone does agree on is that turbo yeast is highly convenient and easy to use.)

Though yeast consumes sugar en route to producing alcohol, sugar does not provide yeast with the fuel it demands to do its job. Yeast runs on nutrients such as potassium, phosphates and nitrogen, and it gets them from grain and fruit sources used in mashes. But in the case of most sugar and water washes, distillers must add yeast nutrients, which are readily available at retail.

**ABOVE:** *Close-up view of the hydrometer as it is being used to test the sugar level of fermenting liquid.*

# WASHES AND MASHES

And speaking of a wash, let's discuss it in detail here. A wash is a mixture of water and liquefied sugar (such as sugarcane juice, molasses or agave nectar) or granulated sugar. Both are dissolved in the water for the purpose of creating alcohol through fermentation. A wash's sugars are immediately accessible to yeast, meaning fermentation begins when it's added.

A mash for grain whiskey is very different. It begins by blending crushed or cracked cereal grains and water. As that mixture is cooked, the grains release starches that gelatinize and turn the mash into a coarse pudding. Next, malted barley is added to the mash and, like magic, it turns the once pasty mash into liquid again. The molecular change is called saccharification, meaning the chains of starches that formerly made up the grain have been broken apart by alpha amylase and beta amylase enzymes in the malt and converted into simple sugars the yeast can now consume. So the mash is cooled, and the yeast is "pitched" in, and the conversion to alcohol begins. If you think fermenting a wash is simpler than a mash, you're right. Without exception, experienced distillers tell me that for beginners, distilling a wash is the easiest way to begin learning the craft.

"We hear people say all the time that they want to get into distilling to make whiskey," said Peter Wheeler, Hauraki BCL, a one-stop distilling supplier in Timaru, New Zealand. "But once they find out how complicated and messy a mash can be, they drop that altogether, start making washes, distilling vodka and back flavoring."

Once you've cooked a mash or warmed a wash, you'll want to take a hydrometer reading before pitching the yeast. This tool, which resembles an old-school glass bulb thermometer, measures the density of a liquid relative to pure water. Since water's density is 1, the density of a liquid mash or wash will be higher because of sugar present in the wash or mash.

For example, if you take a hydrometer of a not-yet-fermented corn whiskey mash, you may get a reading of 1.055. Knowing that number will help you calculate the potential alcohol of your mash, which is usually between 6 percent and 8 percent ABV (alcohol by volume). But just as importantly, knowing that pre-fermentation number will help you determine when your fermentation is finished because your post-fermentation hydrometer reading will return to 0. (We'll delve more into that shortly.)

Back to yeast: following recipes is crucial at every stage of distillation, but perhaps even more so than any during fermentation. Using too much yeast, for example, could produce off flavors in your distillate. Using too little yeast could yield under-fermented wash or mash, which is bad because you won't maximize alcohol production (although if this is the case, the good news is you can always add more yeast to finish the job).

Water temperature is crucial as well. If it's too hot, the yeast will die when pitched. If it's too cold the yeast — not unlike humans — don't like to work. Most recipes for fermenting set the optimal temperature for pitching yeast at about 70°F (21°C), but temperatures do vary slightly with each yeast and recipe, so read the directions carefully.

Once the yeast is pitched, maintaining a consistent wash/mash temperature of 70°F (21°C) is crucial. There are multiple types of thermometers made for temperature measurement of a wash or mash, including those that float in the solution; others are affixed by adhesive to the outside of the fermenter. The main benefit of an exterior thermometer is you never have to remove the fermenter top to check the temperature inside. Yes, the exterior temperature will vary slightly from the actual temperature inside the wash, but experienced distillers say it's not off by enough to make a significant difference in the outcome of your fermentation. What's key here is to store the fermenter in a room where the temperature changes little, if at all.

Once pitched, yeast also must have oxygen to propagate and dominate all microbial activity in the wash. Oxygenating a wash or mash is as simple as stirring the mixture vigorously or pouring it back and forth between containment vessels a few times. As it consumes oxygen and available nutrients, the yeast multiplies by the millions. This stage of fermentation, also called the lag phase, marks a period of aerobic respiration. And when the available oxygen is used up, the yeast will stop multiplying.

**ABOVE:** *Large vats of new whiskey corn mash, fermenting in a distillery.*

Next comes an eating binge for yeast cells, which consume every available sugar molecule in the mash or wash. As they eat, they also excrete carbon dioxide and produce an anaerobic environment. You will know that this is underway when the mixture begins to bubble and form a foamy head, sometimes called the krausen.

Though it is invisible, a thick dome of carbon dioxide forms over the top of the fermenting wash or mash. This is why on large distillery tours, where mash is bubbling in open fermenters, visitors are told not to sniff the ferment directly. Inhaling a large amount of carbon dioxide literally takes the wind out of you.

Whether there is any truth to claims made by some distillers that people have passed out and fallen into the fermentation tank as a result of this, I don't know, but it's both funny and frightening to consider!

In the case of a wash, the solution will become cloudy as fermentation continues. By comparison, a mash is already opaque, but it will turn grayer over time and, in short, become a nasty sight. How anyone centuries ago could examine such a mess and see it as fit for beverage alcohol production is a mystery.

**ABOVE:** *Workers prepare vessels for washing rice prior to the fermentation process at Terada Honke sake brewery in Kozaki, Japan.*

In small-scale fermentation, oxygen must not enter the mash/wash. Were it to reenter the fermentation vessel after carbon dioxide production had ceased, it could oxidize and ruin what is becoming a beer. Therefore, the use of an airlock at the top of any fermentation vessel is essential. This is a valve that allows excess carbon dioxide to exit during fermentation, while prohibiting any entry of oxygen. Held in this oxygen-free state, a distiller's beer can ferment fully and be held safely for a short time before distillation.

As mentioned, yeast is practically magical, and at this point of fermentation, when it's producing alcohol, it's also developing an alcohol tolerance. That's right: it doesn't enter a wash/mash able to tolerate alcohol, it builds its own tolerance. So be wise about your yeast choices by knowing exactly what your alcohol production goal is, what the potential alcohol of your wash/mash is and whether your yeast can tolerate it long enough to ferment all the available sugars.

Depending on your recipe and the yeast you use, fermentation typically takes a few days to a week to complete. Experts say that if your beer isn't fully fermented after that, then there likely are problems. To be certain that fermentation is complete, take a hydrometer reading. If this new reading has returned to 1, its density has decreased because all sugars are "fermented out."

As fermentation ends, the spent yeast will begin to "fall out" of the solution and to the bottom of your holding vessel. This is good, because you want to be able to siphon off the fermented distiller's beer without collecting solids — at least in the case of a wash. A mash always sees some solids carried over to the distillate.

This falling out, however, doesn't always happen on its own. Since there is a great deal of carbon dioxide still suspended in the beer, it can inhibit yeast from falling to the bottom. According to Rick Morris in *The Joy of Home Distilling* (see page 24), you can assist the process by taking the top off of your fermenter and gently agitating the wash with a sanitized paddle or spoon. That agitation releases some of the dissolved carbon dioxide in the beer and clears the way for the yeast to descend. It is likely you will need to repeat this process at least a few times to speed the process along.

And finally to complete the process: racking, which has nothing to do with starting a game of billiards but is the odd name given to the act of siphoning off the distiller's beer. A racking cane — another strange term — is the tool employed for the task of moving liquid from one vessel to another. Tips for this are simple:

- Be sure to sanitize the racking cane before using it.
- Siphon off the beer slowly.
- As the liquid decreases to the point that you near the unwanted solids, do all you can to avoid agitating them to ensure the clearest possible beer.

# CLEANING AND SANITIZING

You can scrub your fermenter until it's odor free and buff your still until it sparkles. But a key step that hobby distillers occasionally skip is sanitizing their equipment thoroughly, and they often learn to regret it.

The main point to remember about still and fermenter hygiene is this: cleaning is not sanitizing and sanitizing is not cleaning. Cleaning removes visible particles and stains from all important surfaces, while sanitizing eliminates bacteria and other microbes invisible to the naked eye. Both must happen to ensure your mashes ferment fully and your distillates don't pick up unintended flavors. Doing both well also extends the life of your equipment.

Cleaning all equipment with soap and water after a run (especially in cases where mashes are boiled with their solids) will help eliminate particulate buildup. Over time, mineral deposits can accumulate and will require some chemical assistance to remove. A variety of professional grade cleansers are available through the distilling resources listed (see page 157) and in your local brewing and winemaking supplies store.

Boiling a 1:1 water-vinegar solution in your still helps, and costs little. Your still size will determine how much solution you'll need; be sure to fill it with enough to allow it to boil for an hour. Once that's finished, dispose of the solution, rinse and scrub with hot water and a non-abrasive brush, then rinse again.

Fermenters should be cleaned and sanitized as soon as possible after transferring the wash or mash to the still. Despite all that sugar being converted to alcohol, the spent yeast and other by-products will still cling firmly to your fermenter if allowed to dry. If you can't clean it immediately, at least rinse it thoroughly to soften any residue and lessen the chance it'll harden.

Veterans say that for ease of cleaning, nothing beats a basic bucket fermenter, whose wide lid provides ample elbow room to get inside and scrub. Avoid using scouring pads as these can scratch the fermenter's surfaces and allow floating particles to stick. Hot and soapy water and a clean rag or large sponge is more than sufficient for the cleaning portion of the job.

Though ever-reliable, the carboy's thin neck requires a bit more finesse to clean, though filling it with hot water and letting it soak for 20 to 30 minutes makes a good start. A flexible, soft-bristled brush is great for cleaning residue from the vessel's neck and shoulders. You just have to be diligent to get it all.

Sanitizers made just for brewing and distilling are widely available, and many are no-rinse products. But Mike Haney, president of Hillbilly Stills, says you can sanitize just as well the old-fashioned way, "All you need is a very small amount of chlorine bleach and water: a 1:20 ratio works best. What you've got to do is get that bleach solution rinsed out well because if there's any left behind, it will kill your yeast in your next fermentation."

Once cleansed and rinsed thoroughly, dry your fermenter — and any tools used — completely with a clean, low-lint towel or paper towels, and store upside down to allow for any missed spots to air-dry. After a half day or so, cover the fermenter and store.

One hobbyist interviewed (off the record) for the book used to be a chef who said the years of scrubbing the kitchen after a long day's work made it automatic for him to clean his still after every run. Not a toothbrush-in-every-connection-point cleaning, he points out, but a good scrubbing, rubbing and drying. "You don't feel like doing it, especially if your day involves a lot of sitting around and sipping what you're making," he says. "But that's part of why you do it with friends, people who don't mind cleaning if they get to drink your liquor. It keeps you from being lazy because once somebody starts cleaning, everybody else joins in and it becomes good fun."

# TROUBLESHOOTING

*ABOVE: The distillate during the double distillation process at a Courvoisier cognac distillery, southwestern France.*

*ABOVE: Smelling samples at The Annual International Degustation of Fruit Distillates in Radimov, Slovakia.*

**Here are several common problems you may experience and their common solutions:**

## DISTILLATE TASTES BAD

The most likely cause is heads, hearts and tails cuts were not made precisely. Another possible issue is your ferment. Stressed yeast produces unpalatable by-products such as esters, fusel oils, diacetyls and polyphenols. As Tripp Stimson (Kentucky Artisan Distillery) says, "If your ferment is bad, then your liquor will be bad, too."

## DISTILLATE TASTES SCORCHED

The solids in your mash stuck to the bottom of your kettle and scorched. Next time, heat your distillate more slowly to allow the solids to become suspended in the mash, or filter them out altogether.

Is your product twice distilled, as in a stripping run followed by a finishing run? Extra runs always produce purer spirits.

## FERMENT SMELLS LIKE VOMIT

Unwanted bacteria has gotten into your mash and caused it to spoil. Dispose of the ruined mash, sanitize your fermenter completely, dry it and start over. If it smells like bad cheese, it is contaminated by a wild yeast called Brettanomyces or "Brett" for short, so start over.

## HANGOVERS

Sometimes these are caused by overconsumption, other times by an improper hearts cut that blended residual methanol from the heads cut. Always be safe, rather than sorry, and collect more heads than you think you need. You can always add them to the next distilling run to capture that desirable ethanol.

## FERMENT IS "STUCK" OR STALLED
**There are lots of potential causes for this, so here's the list:**

### Possibility 1:
### the temperature of your ferment is off.
If it's too hot, yeast will die. If it's too cold, yeast will become dormant and not convert sugar to alcohol. The ideal temperature range for most yeasts (always check the label for specifics) is 70°F to 90°F (21°C to 32°C). You should have a thermometer either inside your fermenter or stuck to its exterior. So the first step is to check the temperature and assess whether that's off.

*If the wash is too hot*, you've likely killed the yeast, but before you discard it you may be able to rescue it by adding some cool water to bring the wash temperature down. Do so without aerating too vigorously, and then wait 24 hours to see if the wash begins fermenting.

*If the wash is too cool*, consider raising the temperature in the room. If that's not possible, move it to a warmer room or provide some source of heat, such as an electric blanket. Just make sure you don't overheat your ferment.

### Possibility 2:
### the wash is too acidic.
This does happen, but pushing the pH below 3 or above 6 is a little difficult to do. First, check your wash with the pH meter. If the pH is below 3, the yeast is at the very least dormant, but likely more likely dead. If you're pH is closer to 3.5, you can add calcium carbonate, potassium carbonate or sodium carbonate to increase pH.

### Possibility 3:
### high starting specific gravity.
If there's too much sugar in your wash, too much ethanol will be produced and it will kill the yeast. Measure with your hydrometer to ensure it is below 1.09. Adding water will lower the specific gravity, but if that doesn't start fermentation, start over.

### Possibility 4:
### too little oxygen in the mash.
Without oxygen, the yeast will not propagate. Try agitating your wash, either with a spoon, a kitchen whisk, or by dumping the contents of your fermenter into an equal-sized container and then back and forth a few times.

### Possibility 5:
### the yeast you pitched is old.
Do your best to remove the old yeast and then pitch new yeast.

## STILL OUTPUT IS SLOW, DRIPS ONLY
It's likely the still temperature is too low. Turn up your heat slightly in steps until a gentle stream begins flowing into your collection vessel. However, watch closely to make sure that the heat increase does not run your wash through too quickly.

## VAPOR, NOT DISTILLATE, COMING OFF STILL
Two issues could be at work here: either the condenser water is not cold enough to condense the vapor into liquid, or the still temperature is too high. In either case, shut down your still immediately to reduce the escape of alcohol vapor into the room.

# VODKA

Vodka is a spirit produced from any arable material distilled above 190°F (88°C) and bottled at no less than 80 proof. Standards of identity vary slightly around the world, but by and large the only additional qualifiers insist that vodka be colorless, odorless and flavor neutral. Yet despite that intended flavor neutrality, were you to taste a half-dozen different brands of vodka side by side, none would taste the same due to the broad range of raw materials — grapes, grains, molasses, tubers, etc. — used in the production of each.

**RIGHT:** *Vodka martinis (see page 48).*

*Vodka is our enemy, so we'll utterly consume it!*

— RUSSIAN PROVERB

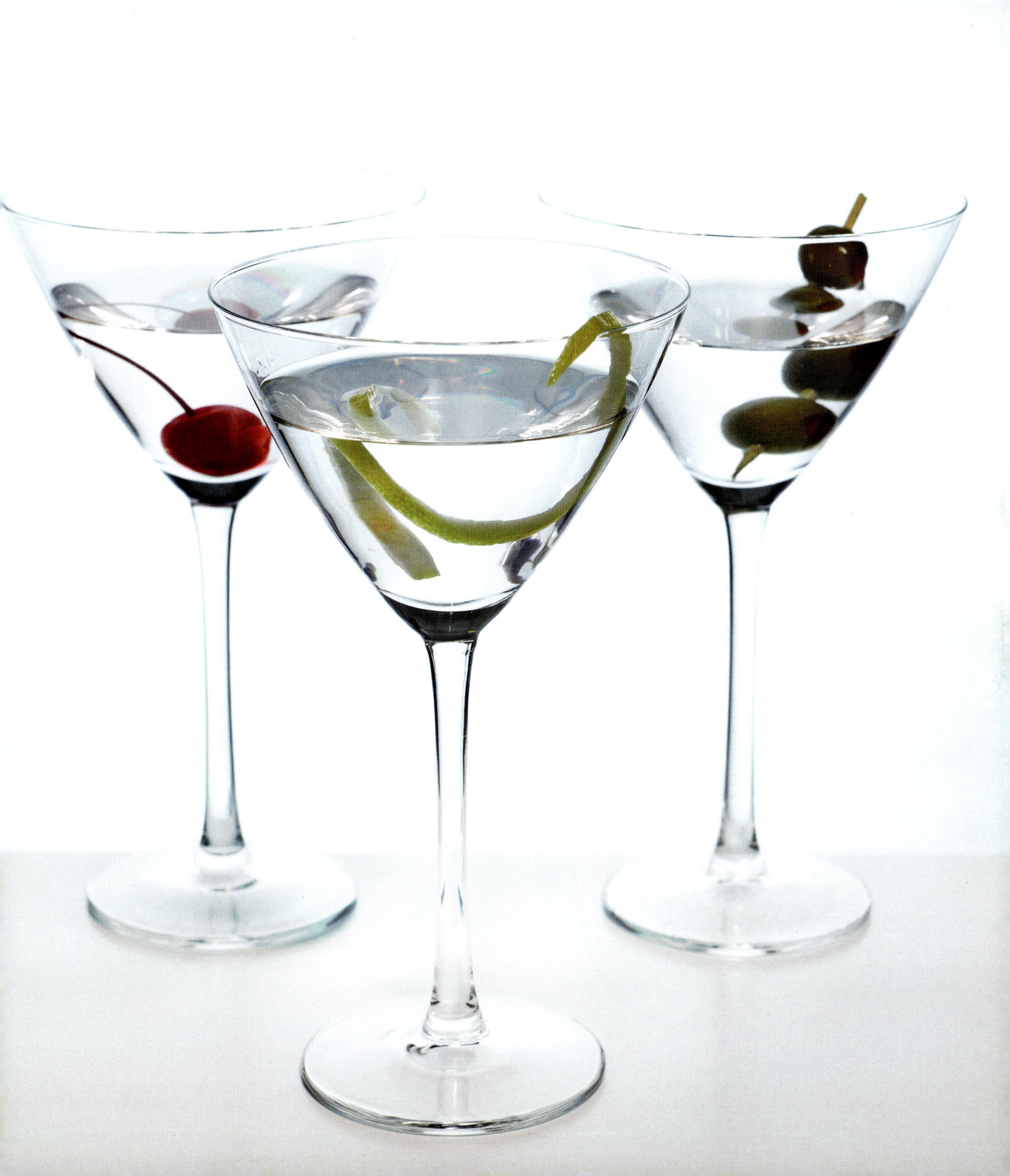

## A SHORT HISTORY

The story behind vodka's beginning isn't quite as clear as the spirit itself. Russia and Poland both lay claim to creating it, but neither can defend their argument beyond the other nation's doubt. What's certain is that distillers in both countries have done a stellar job of producing their versions of this most versatile and adaptable spirit.

Russians claim they started distilling vodka as early as the 12th century, though they concede their products benefitted greatly from the adoption of European distilling techniques (such as the use of rye) and filtration techniques a few centuries later.

In the 16th century, Poland's King Jan Obracht approved the sale of vodka — but only to those who could afford it. A century later, the spirit was sold widely and became Poland's national drink as well as a highly valued export.

Over the ensuing centuries, vodka took on a much broader role due to its versatility as a base spirit redistilled for gin and infused with a wide variety of sweet and spicy flavors. Macerated with fruit and herbs, it was used extensively in liqueurs. Its surging popularity in the United States in the late 1960s and early 1970s slowed sales of American whiskies significantly for many years as drinkers fell for its easy mix-ability in cocktails. Today, consumption of the spirit itself, neat or over ice, is increasing as distillers find cleverer ways to package and market this least flavorful of spirits.

## FERMENTATION

Though vodka is made from a range of fermentable products, for the purpose of simplification, let's stick with one of the most popular of all — potatoes. Understand that if you choose to produce potato vodka from scratch, you are in for a fair amount of work.

To get at fermentable sugars in potatoes, they must be broken down through heating (liquefaction) and with the assistance of enzymes contained in malted grains (saccharification). Unlike a grain mash for whiskey, which is cooked just once in hot water before it's fermented, potatoes must undergo a more time-consuming stepped infusion mash.

For example, a mash of cooked, ground potatoes is added to water, heated to 135°F (57°C) and held at that temperature for 15–20 minutes. (The process includes these pauses because potato starch chains are slow to give up their sugars, and the break in the action allows the enzymes extra time to break down more sugar in each heat-and-stall step.) You then raise the mash temperature to 150°F (66°C) and hold for the same amount of time again. The heat-and-stall cycle is repeated one last time by raising the temperature to 160°F (71°C) and holding yet again. To end the enzymatic reaction, the mash temperature is raised finally to 174°F (79°C) and then cooled as quickly as possible to about 75°F (24°C), when the distiller can begin fermenting it.

Lots of work, right? No wonder that so many distillers interviewed for this book advise hobbyists to source good quality vodka they can redistill and rectify to their own flavor specifications!

## DISTILLATION

In making vodka, the goal is to create a pure spirit. Using a pot still to make vodka will initially fall short of that due to the creation of flavor-imparting congeners. These are eliminated automatically in column distillation (aka reflux still), and they also can be removed in pot distilling through multiple subsequent finishing runs. Using a column still for more precise cuts will leave the vodka much more neutral after just one stripping run and one finishing run.

### The Stripping Run

To begin, siphon off the fermented liquid from your wash and into your still's kettle. Reassemble the still and begin heating the mash gradually to start your stripping run. Heating it slowly will reduce the chance of scorching any residual solids carried over from your fermenter, and lessens the wear and tear delivered from high, direct heat (such as from a propane burner) at the bottom of the still.

Shortly afterward, start running cold water into your condenser. While it may seem early for this, it's a foolproof way to ensure your alcohol will be condensed rather than vented into the air as flammable vapor and

**ABOVE:** *An old vodka factory in Warsaw, Poland. As the Poles increased their distilling skills and production capacity during the 16th century, they began exporting vodka west to the rest of Europe and even east to Russia.*

lost alcohol. If you choose not to start your water so early, at least have a checklist to follow, or set a timer to remind you to turn it on. Additionally, experienced distillers advise you to never leave the room once your still is started, so clear your schedule!

Your initial target temperature will be 174°F (79°C), where methanol begins to boil. Then allow the temperature to rise steadily to a maximum temperature of 207°F (97°C), where the tails cut finishes boiling off. In order to avoid distilling the remaining water in your wash, do not exceed this temperature.

Since this is just a stripping run, where you separate the alcohol from the water in the mash to make a low wine, you won't make cuts. Collect all the alcohol from this run, cool your still, disassemble it, clean it, reassemble it and add the alcohol back to it to begin your finishing run.

## The Finishing Run

Before you begin, take time to label your collection vessels, "Heads," "Hearts" and "Tails," to ensure you do not mix up your cuts. Then start the still and heat the mash gradually. Perhaps the easiest part of making vodka is that you can redistill it if you do mistakenly blend your cuts. If your cuts aren't as precise or the flavor isn't as clean as you'd hoped for, do redistill it.

As your mash heats up and the still begins to run, collect and discard your foreshots and heads. In *The Joy of Home Distilling*, author Rick Morris (see page 24) advises collecting the first 4 ounces (120 ml) of distillate from a 5 gallon (19 l) wash and discarding it. (Obviously, the amount of heads and foreshots increases or decreases based on the size of the wash, so do the math ahead of time.) Hillbilly Stills president, Mike Haney (see page 28), advises discarding the first

**ABOVE:** *Eliminate the handheld hydrometer by using a still parrot, a fixed cylindrical pipe and collection tube welded between the condenser and your final collection vessel. As a temporary reservoir for distillate, it allows for real-time proofing.*

2 percent of liquid emerging from the still (see page 66 for advice on calculating this).

As the heads continue coming off the still, check the proof with an alcohol hydrometer. Ideally, it should be about 80 percent alcohol. When that proof begins to drop, you are distilling hearts. Fill your graduated cylinder halfway and float your hydrometer to watch the proof drop as the still continues to run. Taste the distillate constantly to learn the spirit's flavor at each point. While hydrometers are handy tools for measuring, being able to make cuts using your nose and tongue becomes a point of pride for every distiller.

You also can eliminate the handheld hydrometer by using a still parrot, a fixed cylindrical pipe and collection tube welded between the condenser and your final collection vessel. As condensed distillate flows continually off the still and into the parrot, it accumulates just enough to float the hydrometer. The constant flow of spirits gives the distiller real-time readings on the proof of spirits as they flow off the still. The distillate then overflows out of the parrot and into the final collection vessel.

Continue collecting hearts until the alcohol drops to about 65 percent. Set that aside and begin collecting tails. Distillers do this for two reasons: to use them for flavoring the final distillate; and to redistill later to extract any residual ethanol.

Experienced distillers say your nose is especially good for alerting you to the emergence of tails — because they stink. Some I spoke with cited odors such as "wet dog" and "wet cardboard" as predominant. You can also taste when they emerge from the still: they're unpleasantly rich, unappealingly sweet and can burn the tongue. But when combined in small amounts with

**ABOVE:** *Scott Ervin of NE Minneapolis-based Norseman Distillery cleans bottles before filling with their vodka. The Norseman was one of the first micro-distillers in the area to actually stock their finished product in stores and bars.*

hearts, they can provide flavor notes that give the spirit some character. In the end, the value and beauty of tails is purely in the palate of the distiller.

Once you've collected all the tails, turn off your heat source. Since the remaining liquid in the still is hot, leave your condenser water running to capture residual vapor. Let your still cool before disassembling and cleaning.

## Blending and Storing

If you want to blend some of your tails into your hearts, place all your hearts in a single container and stir in small amounts of tails to suit your personal tastes.

Personal taste also dictates whether you want to purify your vodka further using a carbon filter. Now is the time to do that also. Additionally, this stage is also convenient for "proofing down" your vodka by diluting it with distilled water.

Once you are finished, transfer your vodka to glass or stainless steel storage containers. Or just pour yourself a shot to sip on, or make a cocktail. It's ready to drink!

# HAWAIIAN SIGH

*If there is a story as to how this cocktail was named, it has been forgotten. This is probably because it's a drink that slips down far too easily, so much so that this recipe had to be cobbled together from a multi-person memory bank. Blanco tequila works well in place of vodka.*

## INGREDIENTS
2 oz (60 ml) vodka
1 oz (30 ml) Midori melon liqueur
1 tsp (5 ml) Cointreau or Grand Marnier
½ cup (125 ml) canned pineapple juice
2 maraschino cherries, to garnish
1 pineapple wedge, thumb-sized, to garnish

## METHOD
Pour the vodka and liqueurs into a rocks glass filled with ice and stir for 10 seconds using a bar spoon. Add pineapple juice and blend by pouring into a shaker cup and back into the rocks glass.

To garnish, place the pineapple wedge between two cherries on a skewer.

# IRISH MARTINI

*This chilly sipper appears in many cocktail recipe books. The Irish whiskey rinse gives it a delicious smoky note and a bit of depth that might convince even hardcore gin martini fans to give vodka a chance.*

## INGREDIENTS
2 oz (60 ml) vodka
½ oz (15 ml) dry vermouth
½ oz (15 ml) Irish whiskey
Lemon rind, to garnish

## METHOD
Put the Irish whiskey in a chilled glass and swirl to coat the inside, discarding the excess. Add the vodka and dry vermouth to a small shaker, half-filled with ice. Cover and shake vigorously for about 5 seconds then strain the drink into the whiskey-rinsed glass.

To garnish, peel a strip of lemon rind using a vegetable peeler and fold to express its oils into the glass.

**RIGHT:** *Irish Martini*

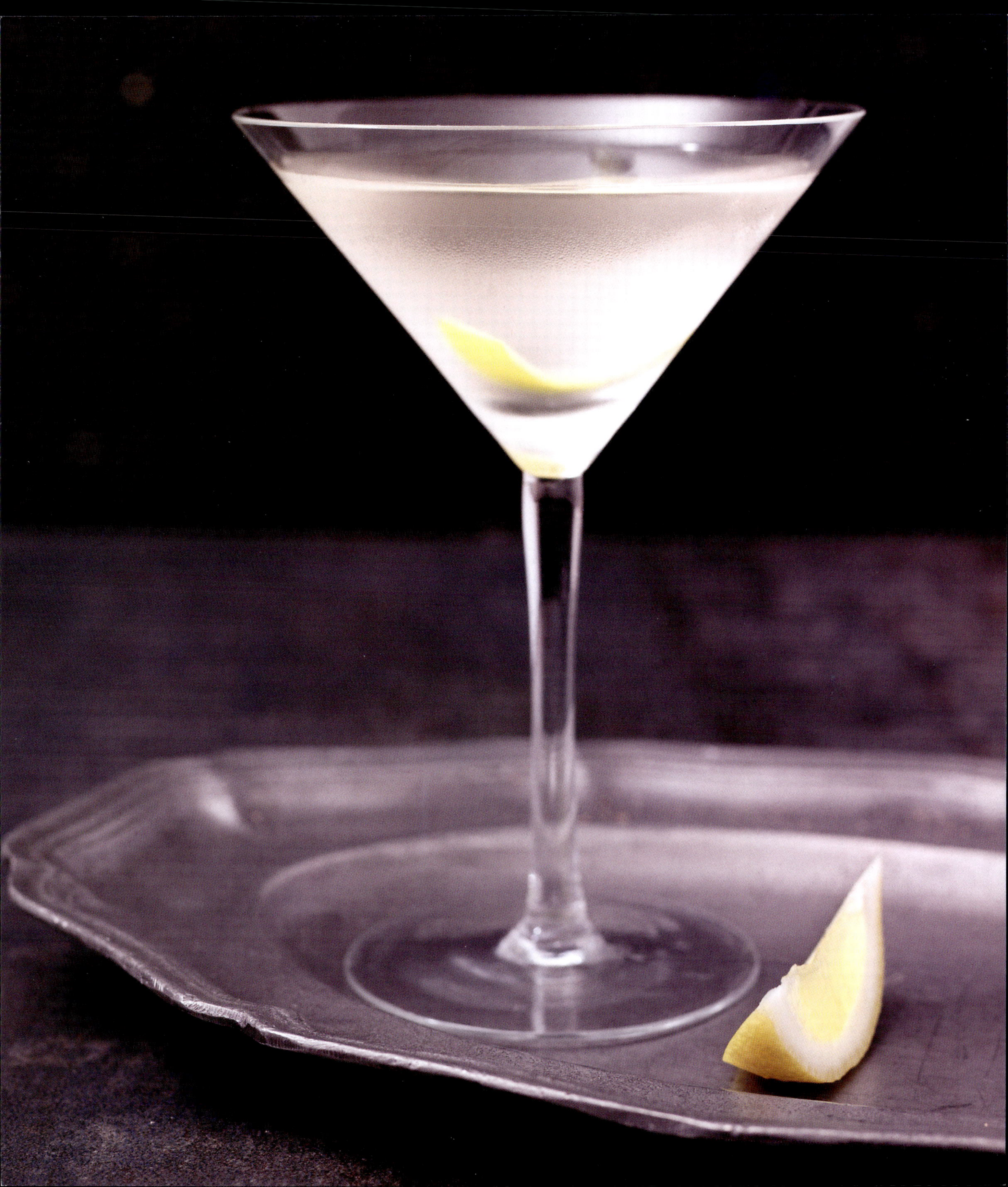

# WORDS OF WISDOM

## PETER WHEELER
### OWNER, HAURAKI BREWING CO.
Timaru, South Island, New Zealand
www.spiritsandbrewing.co.nz

Peter Wheeler wasn't a typical teenager, the type who grumbled about never using his chemistry class lessons in real life. After learning how to distill at school, he and a fellow student launched their spirits-making career.

**Seriously, you really started making alcohol when you were in high school?**

Well, actually, Malcom Willmott and I both started brewing beer when we were 12. In high school we got access to condensers and such, and we figured out how to distill. Of course, it wasn't legal, but it wasn't on a big scale either. Not surprisingly, our neighbors were very encouraging. My mom's only complaint was over the amount of sugar we were using.

**When did it become a business for you?**

About 1989 was when we started to manufacture small home stills. Because of the oil crisis at that time, a lot of people were saying they were using stills to make fuel from crops, so New Zealand repealed the act about still prohibition. When we got into still production, we knew what everyone was going to use it for — human fuel.

What happened in short order was we were freely advertising our stills. Customs would call every few months and ask, "How's business?" And we'd say, "Very good!" Of course they knew what was going on; they were interested, but never did anything. Fortunately, in 1996, they put forward a law that changed it all. Since everyone was allowed to make their own wine and beer at home, they reasoned you should be able to make your own spirits also.

**Did your business boom when it became legal to distill spirits?**

Malcom and I were rubbing our hands thinking, "Oh, we're going to do so well now!" But in reality, everyone who may have wanted a still had one already. So no big sales for us. And since there are only four million people living in New Zealand, we started selling to the

20 million people in Australia. It's not legal to distill alcohol there, but they have a complete disregard for the law, so we shipped a lot of stills to them!

### 🍶 Did legal spirits producers and distributors object?

The Distilled Spirits Manufacturers Association did object. They said the country's going to lose an enormous amount of excise tax, which turned out not to be true at all. Others said the whole country would grind to a halt in an alcoholic haze. I think for a while they were hoping to get some poisonings, though to this day I know of no one being poisoned…well, there was one situation where someone drank some very strong spirit and ended up in hospital. But they happened to be thieves who'd broken into a house and drank something. They didn't get a lot of sympathy.

### 🍶 And some thought distilling would be dangerous at home, right?

Yes, that's true, but it's not been the case at all. None of the stills we make are pressure vessels. The column holes vent at the top, the whole thing is clear; there's no pressure to build up. The worst thing a home distiller can do is not ferment the wash properly and it foams over.

### 🍶 What advice would you give a new hobby distiller?

Buy a distiller package, something ready-made, and ask the person selling it to you for advice. This will cost you around 750 NZ dollars (about $575). Most packages come with DVDs and handbooks to show you how to use the equipment and the key thing is to follow the instructions. The reliability of stills is so good now. Of the many hundreds we produced, we might have had 10 come back, and it typically wasn't due to failure. Usually it was due to someone leaving it in the driveway and backing out over it, or it getting damaged during an argument.

### 🍶 Does size matter?

Yes, it does. I recommend getting an 8-gallon (30-l) still because you get more results for your work.

**ABOVE:** *Two of the 8-gallon (30-l) reflux stills stocked by over 50 retail outlets in New Zealand.*

So many people are thankful we make a 1½ gallon (5 l) still that allows them to not have to distill outside in the shed — they keep it by the couch while watching TV — but they work them to death because they only produce about one bottle at a time, so they're always running them. So why not make larger batches in a bigger still and reduce your workload?

### 🍶 What spirit do you recommend they distill?

Make vodka and then back flavor it. It's simple to make, gives you practice running the still and it's inexpensive and good. A batch of ingredients — sugar, a yeast blend with a haul of nutrients and then your flavoring of choice — will cost about 25 NZ dollars (about $20) per batch, and you'll get five or six 2 pint (1 l) bottles per batch. The flavorings are so good these days: not just chemical concentrates, but real extracts from fruits and other sources. So I say the best way to go is to bang off a spirit and back flavor it.

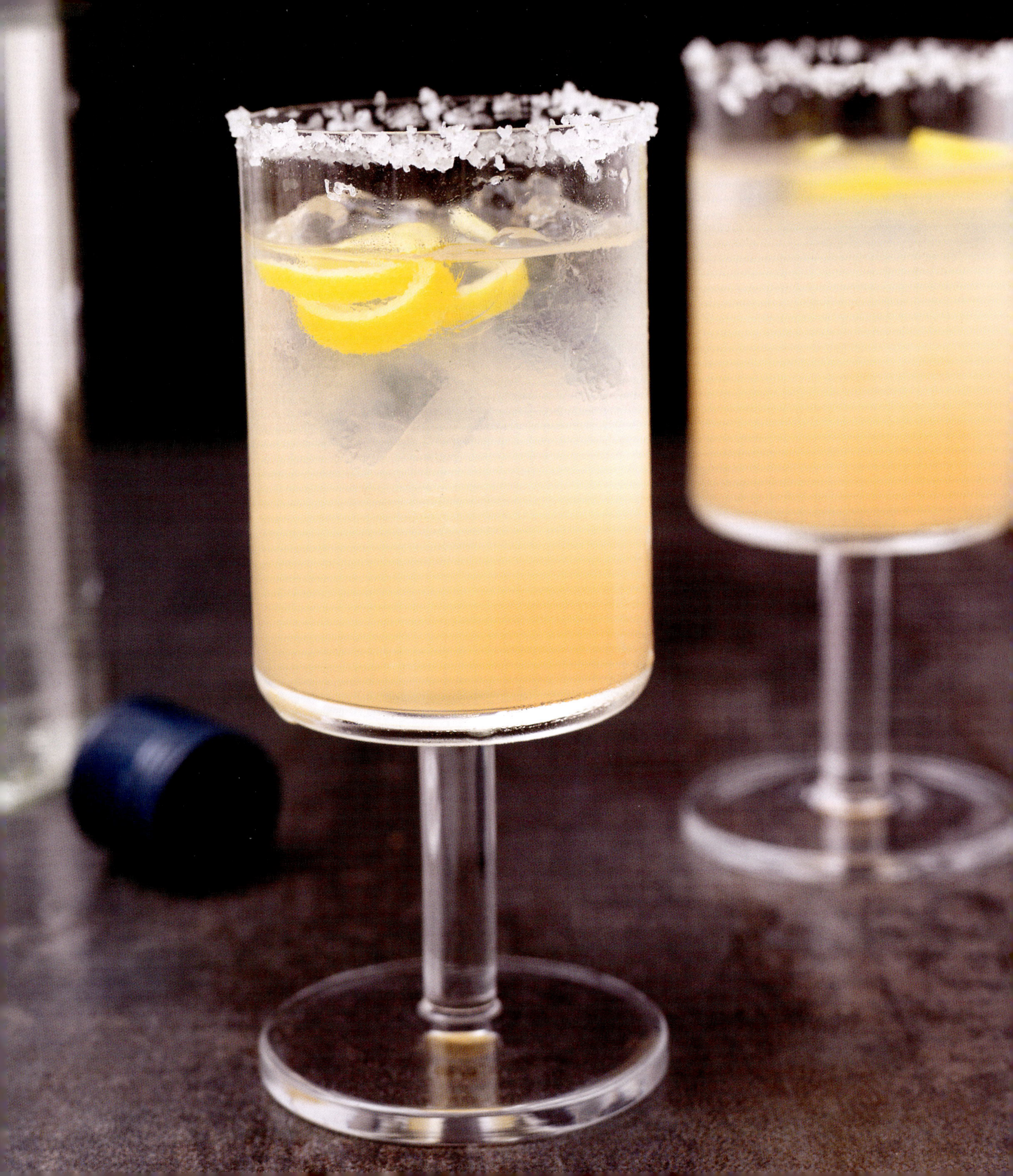

# SALTY DOG

*Multiple sources credit American bartender George Jessel with creating this drink in the early 1950s. That he's also credited with inventing the Bloody Mary can lead one to assume George wasn't a fan of sweet cocktails.*

### INGREDIENTS
2 oz (60 ml) vodka
3 oz (90 ml) white grapefruit juice
Coarse salt
Grapefruit rind, to garnish

### METHOD
To salt rim the glass, pour a small pile of coarse salt onto one small plate and a small amount of grapefruit juice onto another. Dip a rocks glass into the juice and shake off any excess drops; then dip it into the salt, again gently shaking off excess.

Fill the prepared rocks glass with ice cubes, add the vodka and stir for 10 seconds with a bar spoon. Add grapefruit juice and blend by stirring gently.

To garnish, peel a strip of grapefruit rind with a vegetable peeler. Fold to express its oils into the glass.

**LEFT:** *Salty Dog*

# ANGRY HOUND

*The Greyhound is a well-known grapefruit juice and vodka mixed drink, but if you want to give it some pep, add jalapeño peppers and Aperol (a refreshing Italian aperitif). A grapefruit-infused vodka works particularly well here.*

### INGREDIENTS
2 oz (60 ml) vodka
3 oz (90 ml) pink grapefruit juice
1 oz (30 ml) Aperol
1 grapefruit wedge, peeled
2 thin slices jalapeño pepper
Grapefruit rind, to garnish

### METHOD
Put the jalapeño pepper slices and grapefruit wedge in a rocks glass and muddle each lightly (just two twists will do). Add ice cubes, vodka and Aperol, and stir for 10 seconds with a bar spoon.

Add the grapefruit juice, and blend by pouring into a shaker cup and back into the rocks glass.

To garnish, peel a strip of grapefruit rind using a vegetable peeler. Fold to express its oils into the glass.

## THE OUTSIDER

*If you think this drink is named after some tragic misunderstood artist, you'd be wrong — it is named thus because it's perfect for outside consumption in the sun. Though most drinks are best made fresh, this one can be batched and poured from a jug over ice.*

INGREDIENTS
1½ oz (45 ml) vodka
½ oz (15 ml) Pama liqueur
½ oz (15 ml) grapefruit liqueur,
    such as Citron Savage
1 oz (30 ml) freshly-squeezed pink grapefruit juice
1 oz (30 ml) canned pineapple juice
Grapefruit peel, to garnish

METHOD
Half-fill a small shaker with ice and add all the liquid ingredients. Cover and shake vigorously for about 5 seconds and strain into a chilled glass or, if serving outside in hot weather, over ice.

To garnish, peel a strip of grapefruit rind using a vegetable peeler. Fold to express its oils into the glass.

## COSMOPOLITAN

*Sex In the City fairly gets credit for elevating this cocktail, but some bartenders who've researched 1930s recipe books say its lineage can be traced back to the immediate post-Prohibition days when raspberry syrup was used instead of the cranberry juice.*

INGREDIENTS
2 oz (60 ml) vodka
½ oz (15 ml) triple sec, or other orange liqueur
3 oz (90 ml) cranberry juice
1 lime wedge
Orange rind, to garnish

METHOD
Half-fill a small shaker with ice, add the vodka, triple sec and cranberry juice and squeeze in the juice from the lime wedge. Cover and shake vigorously for about 5 seconds, and strain into a chilled glass.

To garnish, peel a strip of orange rind using a vegetable peeler. Fold to express its oils into the glass.

RIGHT: *Cosmopolitan*

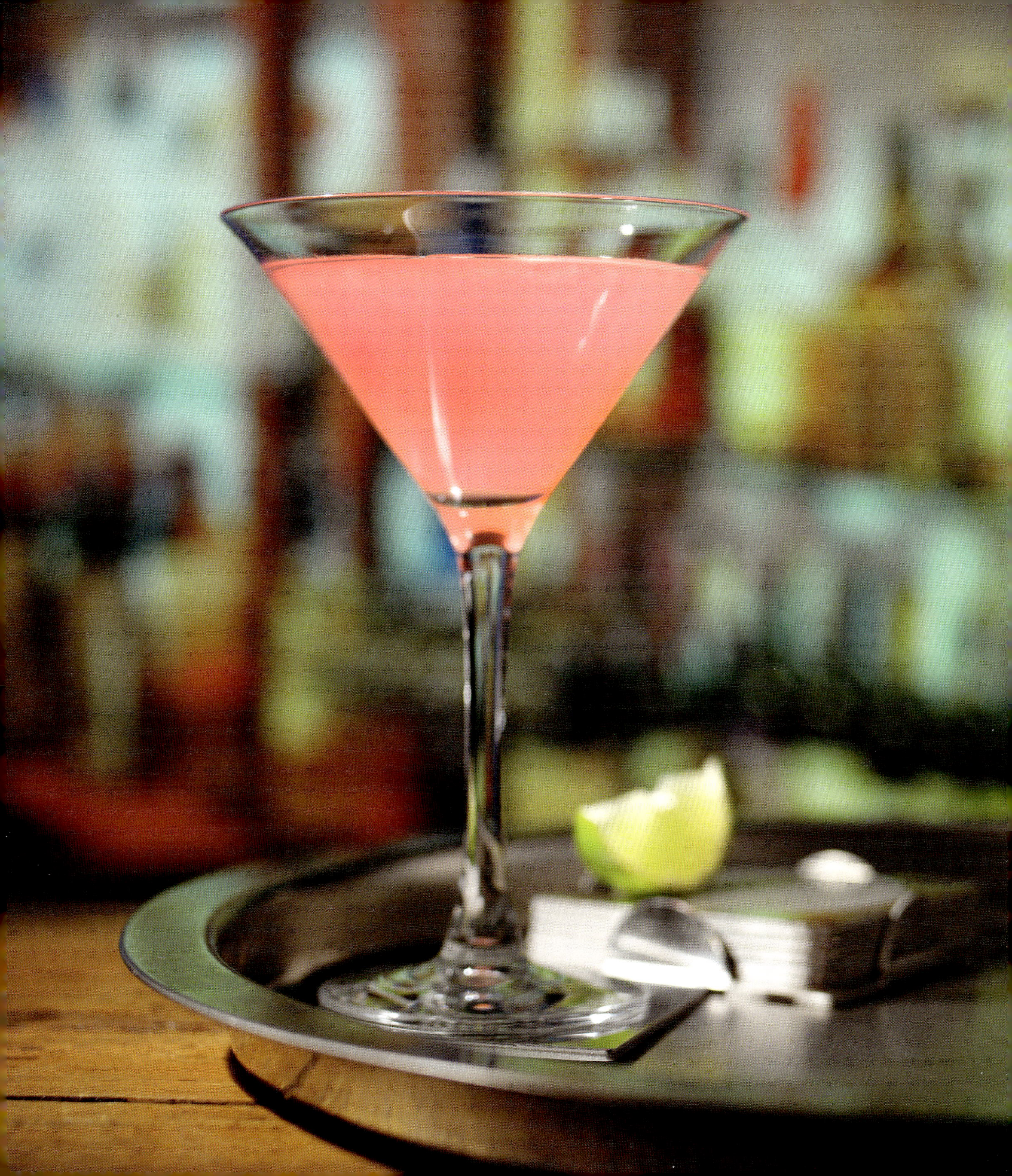

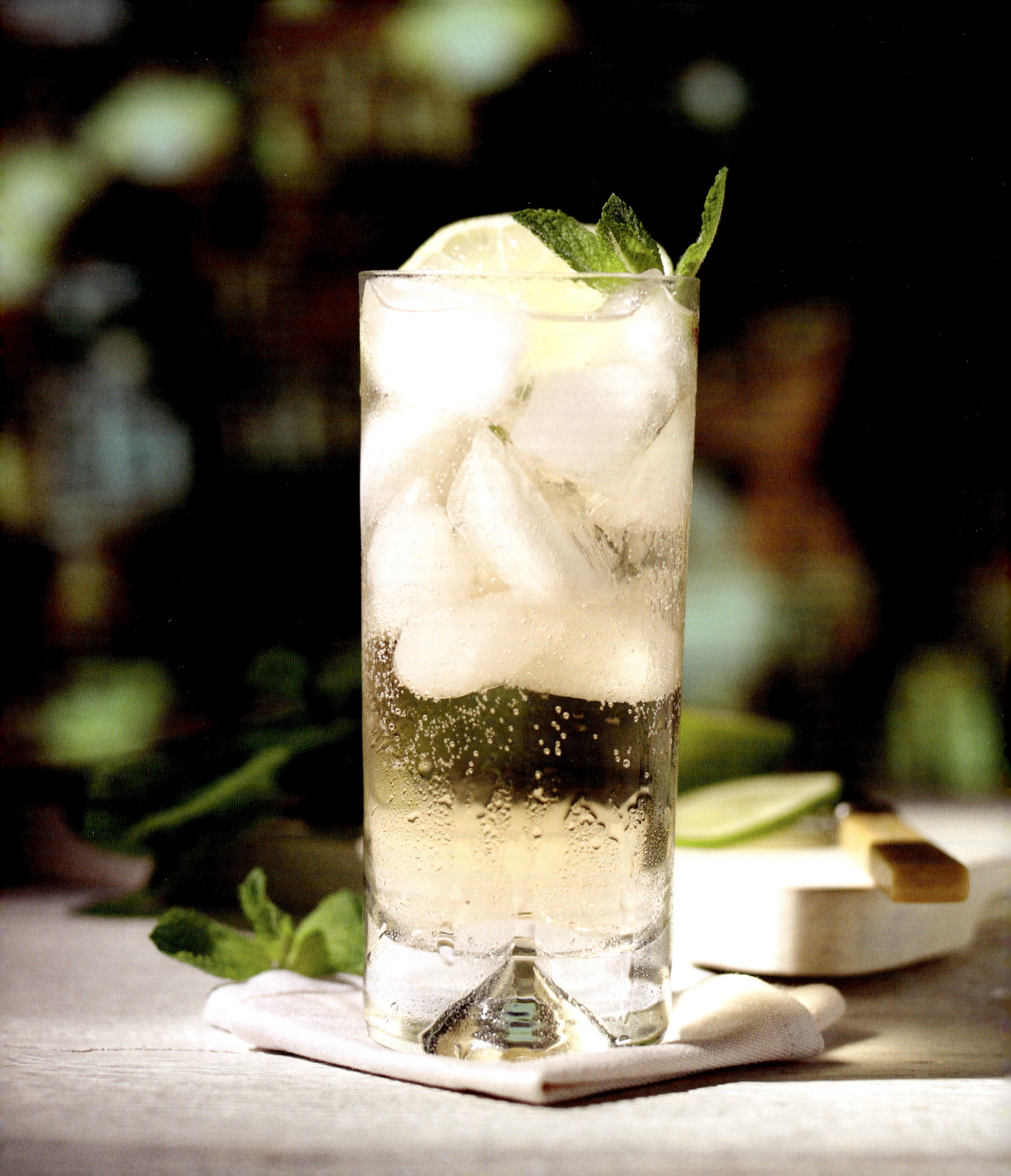

## MOSCOW MULE

*As the story goes, this recipe was created at the Cock 'n' Bull pub in Hollywood in 1946, when a maker of ginger beer, a manufacturer of copper mugs and a vodka salesman came up with an idea to improve sluggish sales of all three products. A copper mug will keep it cold.*

### INGREDIENTS
2 oz (60 ml) vodka
3 oz (90 ml) ginger beer
1 oz (30 ml) freshly-squeezed lime juice (about half a juicy lime)
1 lime wheel, to garnish

### METHOD
Fill a copper Moscow Mule mug or highball glass with ice and add the vodka and lime juice. Using a bar spoon, stir for 10 seconds before adding the ginger beer; stir gently to blend.

To garnish, add the lime wheel.

## THE PROSPECTOR

*A friend of mine created this cocktail and she originally named it Goshen after the town where I lived at the time. Many who liked it hated the name, so she renamed it after the neighboring town — Prospect. It is perfectly finished with a squeeze of lime for a fresh, zesty hint.*

### INGREDIENTS
1½ oz (45 ml) vodka
½ oz (15 ml) orange liqueur
½ oz (15 ml) Pama liqueur
5 oz (150 ml) cranberry juice
Squeeze of lime

### METHOD
Fill a rocks glass with ice, add the vodka and the liqueurs and stir for about 10 seconds using a bar spoon. Add the cranberry juice and a squeeze of lime, and blend by pouring into a shaker cup and back into the rocks glass.

**LEFT:** *Moscow Mule*

# WORDS OF WISDOM

## BRENT GOODIN
### OWNER, BOUNDARY OAK DISTILLERY

Elizabethtown, Kentucky
www.boundaryoakdistillery.com

In learning to make spirits, Brent Goodin never distilled a drop of illegal liquor because he didn't have to. "When you grow up in Kentucky," he says, "the resources are endless, and I got to learn firsthand from the pros."

**So you've never been a moonshiner? Ever?**

No, the first liquor I ever distilled was the day I got my permit in 2013. My grandmother ran the bottling line at Barton's (1972 Distillery, in Bardstown, Kentucky), and she introduced me to real distillers. They taught me and helped me meet others. And let's be real: why make it illegally when you can get really good stuff at any store in Kentucky?

But distilling is also in my genes. My great, great grandfather was one of the original distillers in Kentucky. So when I got into it, it was about heritage. And I was in the right place to do it. We have the raw water source on my family's land outside Elizabethtown; it's a spring that comes out of the ground right by a boundary oak, a tree that marks the property.

**Since you're legal, do hobbyists seek you out for information?**

I'd say half the people who talk to me about distilling are hobbyists. If they aren't, then they know too much about it to not be making it. If they are doing it, I try to give them a few tricks of the trade. I'm not encouraging what they do, but I love sharing what I know.

**Describe these hobby distillers you meet.**

The people who come to us are educated people, highly motivated people who love to cook, love to entertain, love to make things that aren't for sale. They just like to say, "Look what I made!" Honestly, I don't know anyone who wants to sell, but I've heard of some who do. Most though are law-abiding citizens doing it on a small scale.

🍶 **Most of your products are distilled from a sugar wash: why?**

Because corn is a much tougher critter to distill. You have to convert starches in corn through heat, which breaks down those starch chains so they can be converted to sucrose, which the yeast can eat. We have just recently done our first bourbon, but it's a very small batch, and more will follow. But making "sugar 'shine" is also a business decision. I can make it very quickly, get it bottled and on the shelves in a matter of weeks. Those sales keep my business going, and this has always been about business for me.

We're really excited about a product we've called Kentucky Amber, which is our sugar wash distilled to nearly the proof of vodka and watered back down to 105 proof. We take that and store it in used bourbon barrels for 40 days, and it picks up amazing flavor and deep color. My theory is that since it's not a grain whiskey, it has almost no congeners, and that allows the liquid to penetrate the barrel's wood fibers quickly and extract that flavor. Grain whiskey retains lots of congeners that give it body, but again, it's my theory that those congeners slow the penetration of the liquid into the wood.

🍶 **What's the key to good distilling at home?**

Having the proper kind of still and that's a reflux still. I know you can make liquor with any kind of still, but a reflux still is the most effective because you have so much more control than with, say, a pot still. It's true that the more you reflux that liquor, the more flavors will to come out of it, but you get a higher quality alcohol as a result. When you distill through (reflux) plates, you're stripping off congeners you don't want in that drink. But it's not automatic. Even reflux distilling is a dance to determine what you leave in and take out.

🍶 **What's the hardest part of distilling spirits?**

Fermentation. You have to have the right strains of yeast, the right temperature in the fermenter, and your water pH must be correct. And just when you get that down, something will change on you even though you did the same thing as yesterday. You get a lower-proof alcohol or some different flavors.

🍶 **How about some tips for beginners.**

The Treasury Department would tell me to tell people not to do it. But if I were teaching a class, I'd tell people to get a good quality reflux still with a reflux column that will use modern plates. Secondly, don't make a product on a single run: you can do that, but your cuts are going to have to be right on the money, which is really hard, so run it more than once. Lastly, be safe: high-proof alcohol when running through a still has a much higher chance of blowing than low-proof; if you do the right things, that's probably not going to happen, but it can, so just be very careful.

**ABOVE:** *Small batch Kentucky Moonshine 101 Proof is pot stilled using the natural spring fed water of Boundary Oak.*

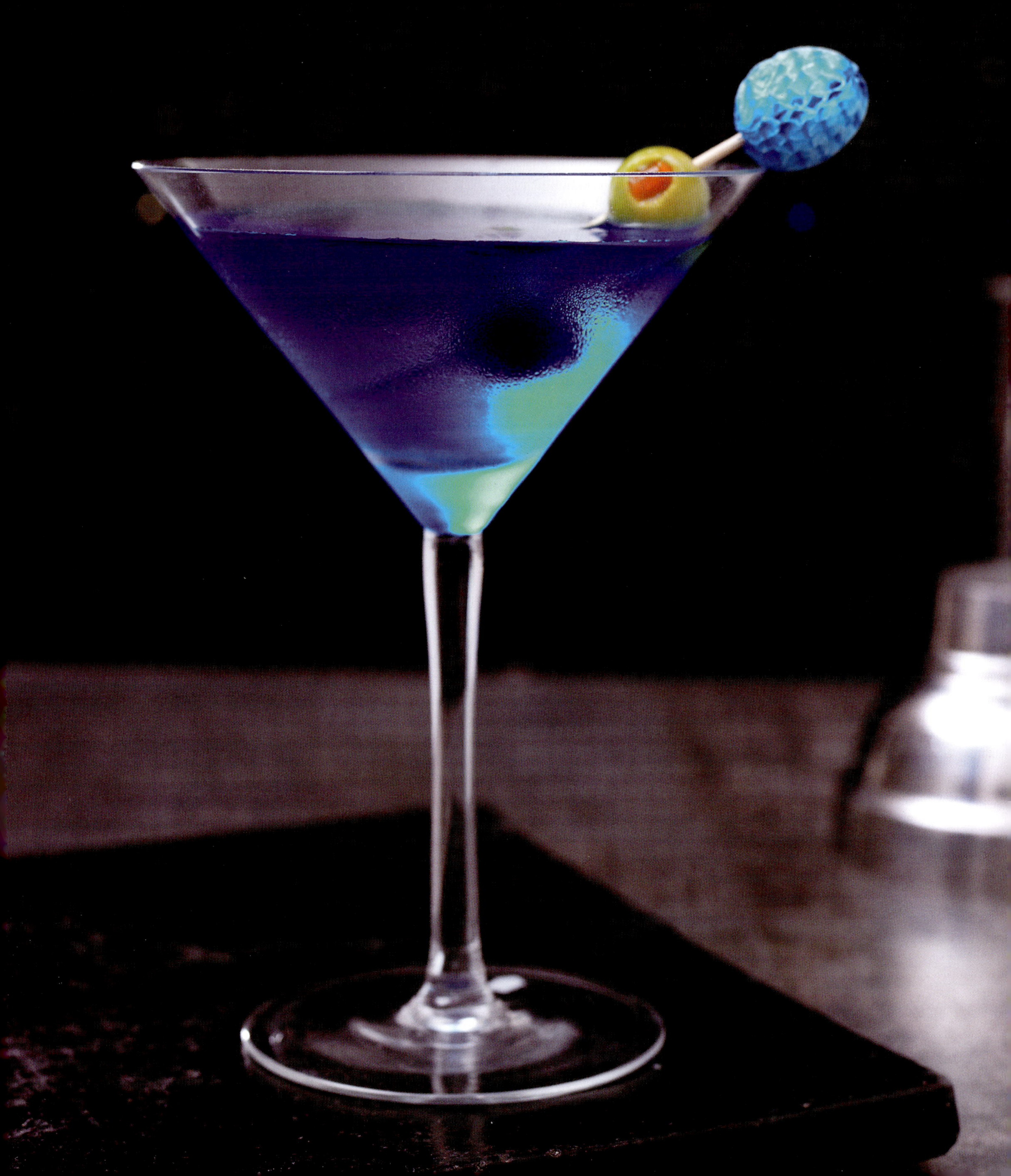

## WILD BLUE YONDER

*While no history can be found for the naming of this cocktail, it's likely that after sipping one of these, you will feel like staring blissfully at a clear, blue sky. If you're not a fan of the salty olive contrast, you could try garnishing with a skewered orange slice.*

INGREDIENTS

2 oz (60 ml) vodka
1 oz (30 ml) sweet vermouth
½ oz (15 ml) Blue Curacao
2 olives, to garnish

METHOD

Half-fill a small shaker cup with ice, add the vodka, sweet vermouth and Blue Curacao and stir for about 20 seconds. Strain the drink into a chilled glass and garnish with the skewered olives.

LEFT: *Wild Blue Yonder*

## GROUP HUG PUNCH

*Punch is much more fun to make when it's a group project, and a potent mix like this tends to bring out the love between the punch makers. It's best served chilled, so make it up 6 to 12 hours in advance and keep it in a sealable container before transferring it to a glass pitcher.*

INGREDIENTS

2 cups (500 ml) vodka (or infused orange or raspberry vodka)
2 tbsp (30 ml) orange liqueur
4 cups (1 l) cranberry juice cocktail
2 cups (500 ml) freshly-squeezed orange juice
2 tbsp (30 ml) agave syrup
½ tsp ground cinnamon
¼ tsp ground cloves
1 mint stem with about 12 leaves
10 thin half orange slices, to garnish
20 thin half lime slices, to garnish
Cranberry juice ice (optional)

METHOD

Blend the liquids, ground cinnamon and cloves in a large, sealable container. Tie butcher's twine to the end of the mint stem and immerse it in the cocktail blend. Seal and refrigerate for 6 hours before discarding mint.

Place 1 orange and 2 lime slices into the bottom of each rocks glass. Fill halfway with ice and add 1 or 2 cubes of cranberry juice ice (optional). Top up with punch.

# WHISKEY

Whiskey is a distilled spirit made from cooked and fermented cereal grains. Too bad it's not quite as easy to make a good whiskey as it is to define it. It's not rocket science — after all it's been distilled by the uneducated masses in crude stills and under less-than-ideal circumstances for well over half a millennium — yet the difference between poorly made and well-made whiskey is where the real expertise lies; it requires practice, careful preparation and attention to detail. Every part of the process, from fermentation to aging, can be improved with time and experience. No wonder then that some of the finest whiskey distillers the world 'round are centuries old.

RIGHT: *Whiskey shots — best served neat.*

*Always carry a flagon of whiskey in case of snakebite, and furthermore, always carry a small snake.*

— W.C. FIELDS

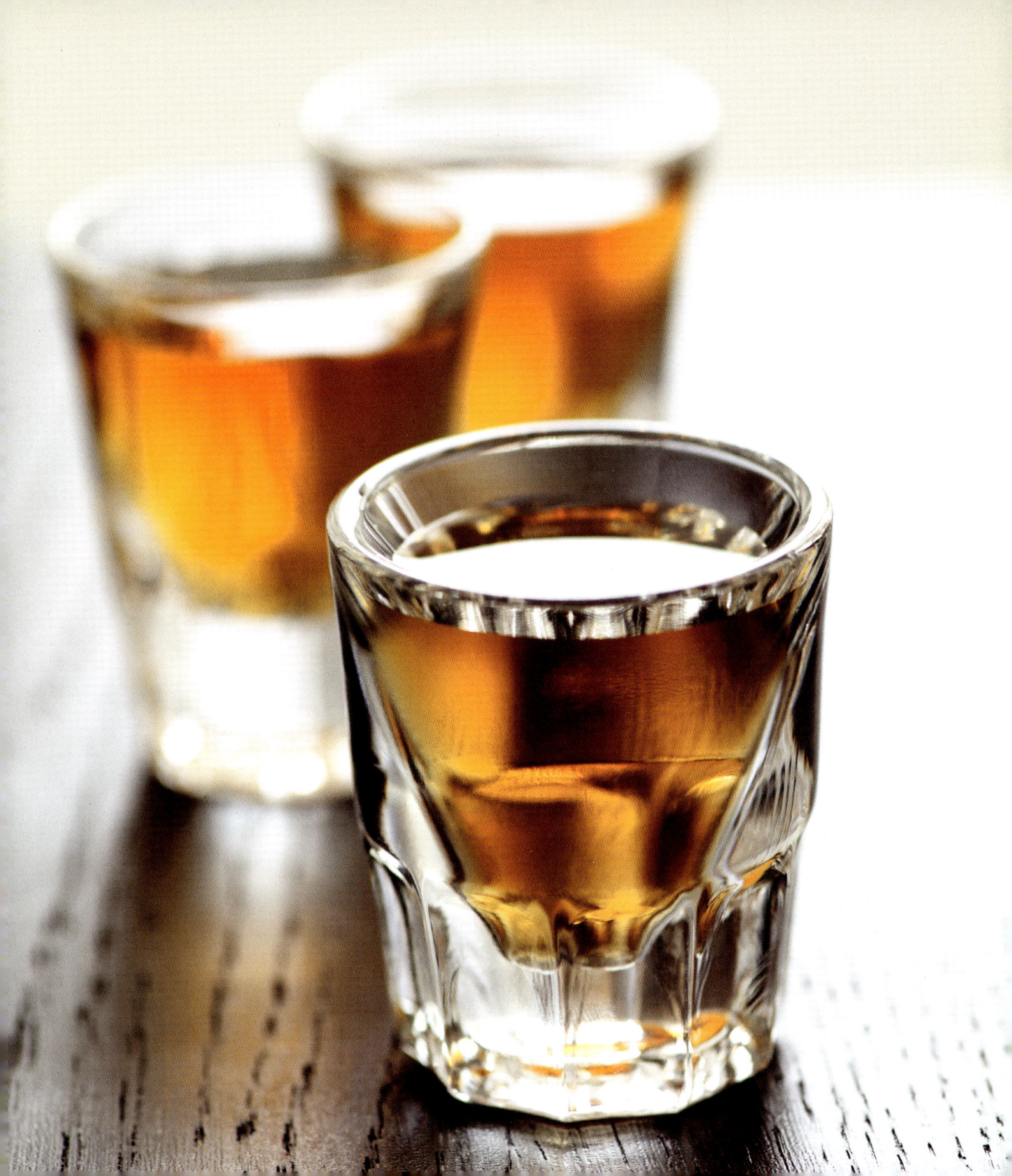

## TYPES OF WHISKEY

All whiskies are made following highly similar methods, but where they differ in their nuances makes a world of difference in their flavors. So let's take a quick look at the major types based on how they're defined in their country of origin:

### Bourbon

Whiskey produced in the U.S. not exceeding 80 percent ABV (160 proof) from a fermented mash of at least 51 percent corn and stored at not more than 62.5 percent ABV (125 proof) in new, charred American-oak containers.

### Irish

Whiskey distilled in Ireland from a fermented mash of mixed grains at an ABV of less than 94.8 percent and aged in wooden casks for at least three years.

### Scotch

Whiskey distilled in Scotland from a fermented mash of barley to an alcoholic strength by volume of less than 94.8 percent; aged in wooden casks for no less than three years and to which nothing other than water or spirit caramel is added. Scotches are bottled as blends (sometimes a blend of distillates, other times a blend of malts) or as single malts (made at one distillery only and in pot stills).

### Canadian

The only requirement for Canadian whiskey is the distillate must be aged in wood containers smaller than 700 liters (185 gallons) for three years and be bottled at 40 percent ABV (80 proof). Canadian whiskies commonly are blends of distillates made from separate mashes of multiple grains.

Each country's rules on single-grain distillates such as rye or wheat may vary slightly, but they largely adhere to the same standards as their country's most prominent whiskies. With home distilling you can, if you choose, rewrite the rules as you see fit because you're not selling it to fit a commercial standard.

## COOKING

Before you begin, create a checklist of all you'll need for each step. For cooking and fermenting you'll need:

- Cracked or milled grains
- Enzymes
- Large cooking pot or kettle
- Heat source (stove top or electric hotplate)
- High-temperature thermometer
- Hydrometer
- Graduated cylinder
- Fermentation bucket
- Secondary bucket for aerating
- Kitchen whisk
- Stirring spoon
- Yeast
- Strainer and cheesecloth, if making a mash

### Making a Mash

Following your recipe, heat water to 160°F (71°C), stirring regularly to avoid scorching. Turn off the heat, stir in milled or cracked grain and let the grain soak for 1 hour to hydrate. Add enzymes and heat the mixture gradually to 155°F (68°C); hold for 45 minutes, stirring regularly to avoid scorching. The mash will now be quite liquid. Once cooked, chill quickly to 90°F (32°C) (see page 65) and ferment.

### Making a Wort

Heat water to 160°F (71°C), add milled/cracked grain and soak for 1 hour. Add enzymes and gradually heat to 149°F (65°C); hold for 45 minutes. Stir to avoid scorching. Piece together a collection pot at least the volume of the cooking pot's contents and place a large fine-sieve strainer (or standard cooking strainer layered with several sheets of cheesecloth) on top.

When cooking is complete, pour grains and liquid into a strainer. Take about one-quarter of the captured liquid, heat to 160°F (71°C) and pour that water slowly over the grain remaining in your strainer. Remove one-quarter of that liquid, return to the stove and heat to 180°F (83°C). Pour that liquid over the cooked grains again. Once thoroughly drained, discard the grains and cool the liquid as quickly as possible.

**ABOVE:** *Kentucky whiskey corn mash bubbling away in distillery vats.*

**ABOVE:** *Tripp Stimson of Kentucky Artisan Distillery, who is credited with creating their first bourbon, Whiskey Row.*

### Cooling Your Mash or Wort

If you are producing a small batch of a few gallons, fill your sink with ice-cold water and place your pot inside the sink. Stir with a spoon until cool. Cooling quickly helps to reduce unwanted bacteria forming.

## FERMENTATION

Kentucky Artisan Distillery's Tripp Stimson, originally a chemical engineer, stresses that "Good fermentation is key. Distillation will not make up for bad fermentation, so make sure you do this right."

Once your mash (or wort) is cooled to 90°F (32°C), you'll want to determine your potential alcohol. So take your graduated cylinder, fill it to about two-thirds full, then float your hydrometer in the liquid. This will tell you how much fermentable sugar exists by determining specific gravity or brix. For example, if the hydrometer measures a specific gravity of 1.055 or 13.7 brix, your mash has a potential alcohol of 7.2 percent. That means if you have a 3 gallon (11 l) wort or mash, it would potentially yield just over 27 ounces (about a 750 ml bottle's worth) of alcohol. A good rule of thumb for potential alcohol in a wort or mash is 6 to 10 percent.

With your mash cooled to 90°F (32°C), aerate the liquid by stirring gently with a large whisk or dumping the mixture back and forth a few times between your fermentation bucket and cooking pot. Do not over-agitate with the whisk or dump the liquid more than a few times. Adding too much oxygen can interfere with the formation of esters, which provide flavor. Once aerated, pitch in the yeast and blend with a spoon.

Almost immediately, the yeast cells will begin multiplying and consuming oxygen. Once that's used up, it will begin consuming all available sugars and producing carbon dioxide and alcohol. This will be visible through bubbling in the mash. Because fermentation is an exothermic activity, your ferment will become warm. Monitor the thermometer on your container regularly. Don't let the temperature exceed 95°F (35°C), or you'll risk killing the yeast. If the temperature slips below 80°F (27°C), yeast activity will slow and your mash will not be fully fermented.

## DISTILLATION

Before you begin, check your list of equipment:
- Heat source
- Still
- Alcohol hydrometer
- Three collection vessels for heads, hearts and tails cuts

### The Stripping Run

Pour the fermented mash into the still. Though some distillers also transfer residual solids from the ferment to the still, that's not recommended for hobbyists until they have a few solid runs under their belts. Just as cooking can scorch the grains, distilling can scorch any solids in the kettle. So, until you've mastered your heat source, sidestep the temptation to add any solids.

Fill the still two-thirds full. The extra space allows for proper vapor travel, reducing the threat of foaming or clogging the column or condenser. Once the still is secured, heat it slowly, gradually increasing your heat. In this initial stripping run, you just want to get the alcohol out of the mash and into the condenser. Don't worry about heads or tails cuts at this stage.

As your still heats up, ensure the condenser is getting plenty of cold water. Though it may take an hour or more to see distillate run, your condenser should always be ready to cool the alcohol vapors as soon as they emerge. If any vapors are coming from your condenser, shut down the still immediately. Alcohol vapors are flammable and can pose a fire risk.

Once alcohol begins to trickle out of the condenser, collect it into a glass or stainless steel vessel, fill half your graduated cylinder with alcohol and float your alcohol hydrometer in it to gauge your distillate's proof. This distillate is called "low wines" and ranges between 45 proof and 50 proof. When that proof begins to decline, you're likely starting to condense water, which you don't want. So shut the still off and set your alcohol distillate aside. Once fully cooled, disassemble the still to discard the water and any residual ferment. Once cleaned, add your stripped distillate and reassemble.

### The Finishing Run

This is when you separate good alcohol from bad and delicious flavors from awful ones. Don't leave the still unattended at this time. Once the alcohol begins to run, which will happen more quickly than on the stripping run, your full attention will be required. And this is the fun part that you don't want to miss.

The first distillate to come off the condenser will be the foreshots or the heads cut. This is mostly methanol and must be discarded for the sake of safety. It is pure poison. Mike Haney, president of Hillbilly Stills, recommends discarding the first 2 percent of alcohol that comes off the still to ensure you have captured all the heads. As Haney says, "Even if you throw away a couple of ounces extra, your liquor will be that much safer and that much better tasting. It's better to be safe than sorry."

Though you do not want to drink the foreshots or the heads cut, it won't hurt to taste them: it's a good way to learn to tell the difference between the heads cut and the hearts cut. Haney recommends using fingertips to taste cuts since they are neutral, relative to your own palate. So as the distillate runs, sample it regularly to learn how each portion of the run tastes.

Once you are into the hearts cut, you'll notice a distinct and positive change in the taste and smell of the distillate. Since it's nearly all ethanol coming off the still, it will have a lean texture that is largely free of heavier fusel oils. During this portion of the run, collect some of the spirits in your graduated cylinder and check them with the alcohol hydrometer. According to Bryan Davis, author of *How to Make Whiskey,* your distillation will be complete when the hydrometer reads somewhere between 52 percent and 72 percent alcohol. Most whiskies, he writes, will favor the high end of that range.

> **CALCULATING 2 PERCENT: AN EXAMPLE**
> If the stripping run yield is 5 gallons (19 l), your finished yield will be about 75 percent of that, or 3¾ gallons (14 l), and 2 percent of that is about 9½ ounces (280 ml).

**ABOVE:** *Whiskey barrel house, where the barrels are well ventilated and the whiskey inside is able to age in both airtight and watertight conditions.*

As your hearts cut ends, the tails will follow immediately. Tails can be useful to the flavor of your final distillate since they add sweetness and body. But since a little goes a long way, it's wise to use separate collection vessels for tails and hearts. That allows you to add back as much as you want from your tails cut without overwhelming your hearts cut.

As with any distillation run, take copious notes. Experienced distillers take regular readings with an alcohol hydrometer and jot down tasting notes beside each number. For example, if the distiller likes the taste at 65 percent alcohol, he should note that. Then, as the amount of alcohol coming off the still begins to decline, signaling that the tails cut is beginning, he can note that as well and decide whether the taste is favorable. Making detailed notes is a great way of learning to make precise cuts. Eventually, you will learn to rely solely on your palate to make those cuts.

## STORAGE AND AGING

High-proof distillate should be kept in sanitized glass or stainless steel, as neither is affected negatively by alcohol. Plastic containers can sometimes be damaged and can leave an off flavor in whiskies.

If you plan to store whiskey in a wooden barrel, be sure to test it beforehand: fill it with water to check for leaks and rinse it remove excessive build-up of char dust. Fill the barrel as completely as possible to ensure maximum contact with all charred surfaces. Then, as Tripp Stimson (Kentucky Artisan Distillery) advises, move the barrel to a location subject to temperature changes, such as a ventilated shed or garage. Once shelved, leave it to age without agitation.

When aging grain whiskey, changes in color and flavor happen slowly. So every few months, remove some whiskey from the barrel to taste it and assess its color, taking notes for future reference.

# OLD FASHIONED

*Legend has it that the delicious Old Fashioned was created at the stuffy old Pendennis Club in Louisville, Kentucky. Wherever it was born, we should be grateful to the creator of this simple, yet potent, cocktail.*

## INGREDIENTS
1½ oz (45 ml) bourbon (or rye whiskey, or a blend of half bourbon and half rye whiskey)
2 dashes Angostura bitters
1 sugar cube
1–2 dashes water
1 maraschino cherry, to garnish
Orange rind, to garnish

## METHOD
Place the sugar cube into a rocks glass, saturate with bitters and a dash or two of water, and muddle until dissolved. Half-fill the glass with ice cubes, add whiskey and stir for about 10 seconds using a bar spoon.

To garnish, peel a strip of orange rind using a vegetable peeler and fold to express its oils into the glass. Add a maraschino cherry and serve.

# MANHATTAN

*The name solves any mystery as to where this "whiskey martini" was created. And even if you're a gin purist, you should try to "get down with the brown" with this sipper. Its simplicity screams perfection.*

## INGREDIENTS
2 oz (60 ml) rye whiskey (high-proof bourbon is a delicious substitute)
½–1 oz (15–30 ml) sweet vermouth
2–3 dashes of Angostura, Peychaud's or Regan's Orange bitters
1 maraschino cherry, to garnish

## METHOD
Half-fill a small shaker with ice, add the rye whiskey, sweet vermouth and bitters and stir for about 15 seconds. Put the cherry into a chilled glass and strain the liquid over. Not cold enough? Serve it on the rocks!

# BLOOD AND SAND

*Let's be honest, this drink sounds more like a tribute to the troops who gave their lives on the beaches of Normandy than a desirable evening sipper. However, it is one of the classiest of all the cocktails devoted to their beloved single malts.*

## INGREDIENTS
¾ oz (22 ml) good single malt Scotch
¾ oz (22 ml) sweet vermouth
¾ oz (22 ml) Cherry Heering cherry liqueur
¾ oz (22 ml) orange juice
Orange rind, to garnish

## METHOD
Fill a small shaker with ice, add all the liquid ingredients, cover and shake vigorously for about 5 seconds. Strain into a chilled rocks glass or, to class it up, a chilled coupe glass.

To garnish, peel a strip of orange rind using a vegetable peeler. Fold to express its oils into the glass.

# THE SAZERAC

*New Orleans pharmacist Antoine Peychaud is credited with creating this outstanding drink in the 1830s, although surely it was never hard to get this medicine down! Buffalo Trace Distillery makes two straight ryes under the Sazerac name, and they're both excellent.*

## INGREDIENTS
1½ oz (45 ml) Sazerac rye whiskey
   (or any high rye bourbon)
¼ oz (7 ml) Herbsaint liqueur
3 dashes Peychaud's bitters
1 sugar cube
Lemon rind, to garnish

## METHOD
Take two rocks glasses: fill one glass with ice; place the sugar cube in the second glass, saturate it with the bitters, muddle until blended, and then add the whiskey. Empty the ice from the first glass and add the Herbsaint; roll the liqueur around the inside of the glass to coat it and discard the excess. Tip the whiskey mixture into the Herbsaint-coated glass.

To garnish, peel a strip of lemon rind using a vegetable peeler. Fold to express its oils into the glass.

**RIGHT:** *The Sazerac*

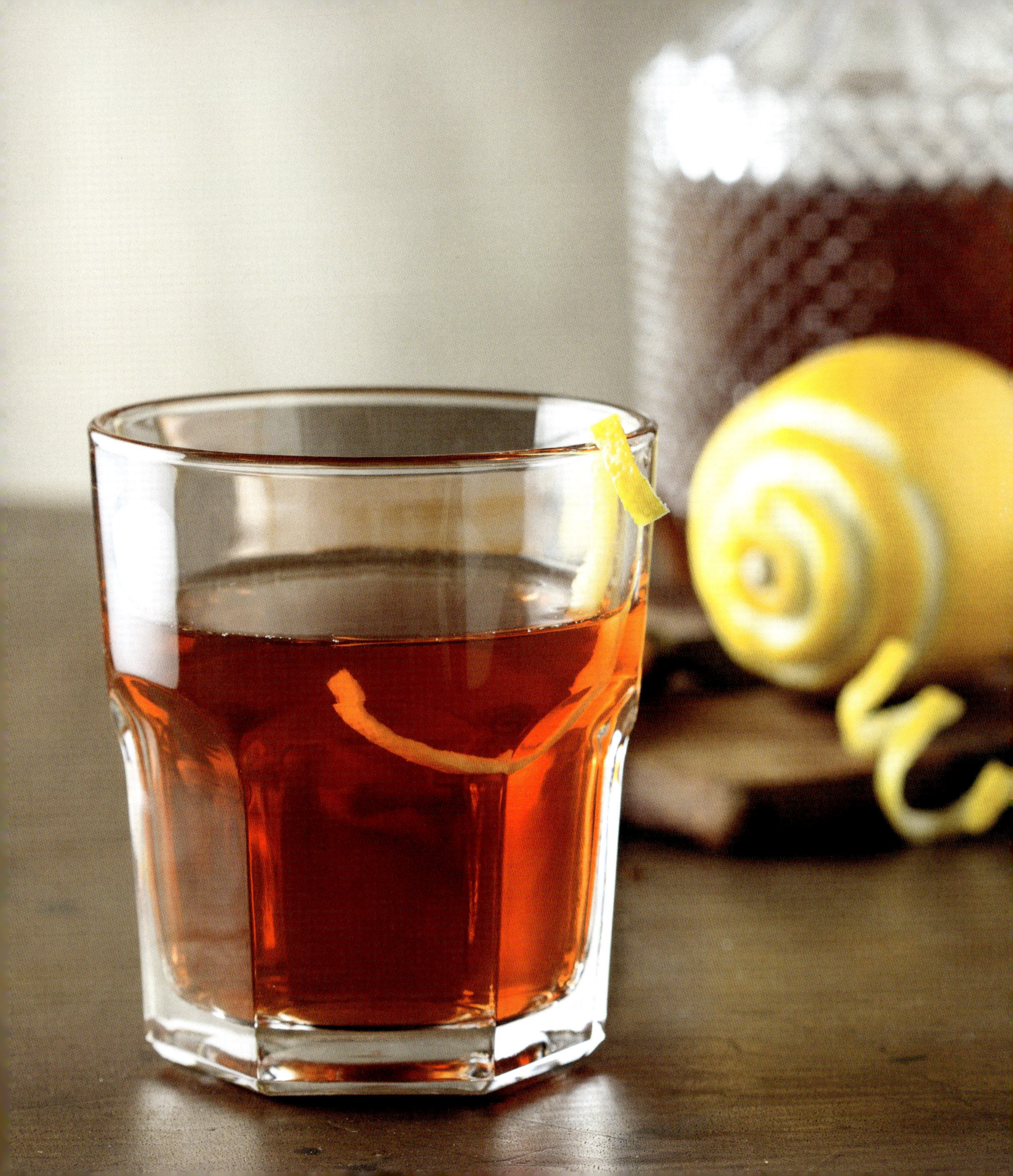

# WORDS OF WISDOM

## BARRY BERNSTEIN

**Partner, Still Waters Distillery**
Concord, Ontario, Canada
www.stillwatersdistillery.com

It wasn't a midlife crisis that led Barry Bernstein and Barry Stein to cofound Still Waters Distillery in 2009; it was their shared desire to finish their working lives having fun. The two had done well in other industries and, with a now-or-never drive, they began making Stalk & Barrel brand single-malt whiskey, vodka and rye.

Distilling that good liquor, Bernstein says, is actually the easiest part of the business. Marketing and brand building has been their biggest challenge.

### So you never were home distillers?

No, that's asking for trouble. We looked at what was happening in the United States with the proliferation of micro-distillers and didn't see many in Canada. So we saw that as a great opportunity. And given Canada's rich history in distilling, we wanted to get involved. Since we grow some of the best grains in the world here, why not make some of the best whiskeys in the world?

### What did you do before opening the distillery?

I had a software business and my partner worked for a large multinational. We were at a point in our lives when we thought we should do something we love, and since we loved whiskey, we decided to go for it. It's good that we had done fairly well before, because this is a business that requires an awful lot of capital.

### I understand that the Canadian liquor industry is highly regulated.

There's a lot of attention paid to what you're doing. It would be pretty hard to get away with something, but from the home distilling perspective, I'm sure that it goes on all the time. We've met a lot of people who have commented on our processes and said, "Oh, yeah, I'm familiar with it, I do that at home." That's when I say, "Don't tell us that! It's not legal and we don't want to know!"

### How difficult was it to obtain a license to distill there?

It's not expensive, really, to get a license that allows you to produce alcohol. But to get the license, you have to invest in your facility and equipment and make the business case saying you'll run the business in a responsible way.

🥃 **You had to have all of your equipment in place before you started?**

That's right. To show them we had a legitimate business, we had to be ready to distill. They will not issue a license to you if you can't convince them you're prepared to operate it professionally.

Here's another thing that's different from the United States: To be called whiskey in Canada, it has to be aged for a minimum of three years. So there's no "new make" called whiskey, no six-month-old brown spirits called whiskey that you will find on a shelf. So when you see that's three years of no cash flow, it makes a lot of people think twice about opening up distilleries here.

🥃 **Since you did not distill at home, how did you learn the craft?**

A lot of text book knowledge and we did take a course. But when you get your hands into it, that's when you really learn. Distilling is not a difficult thing to do from a technology standpoint, but the business aspect has been the hardest part. Licensing and getting product to market is difficult in Canada, and small distillers like us are looking for advice on the business side.

🥃 **Was it so difficult that you wish you'd not gotten into it?**

No, but when people ask us about starting a distillery, we tell them, "Don't do it here. Move down to the United States and do it." It's very hard here, especially in Ontario, where the province runs the sale of alcohol and taxation is extraordinarily high. Still, we really enjoy making whiskey.

🥃 **So let's talk about those single malts you've made.**

What we wanted to do was produce a world-class single malt whiskey made in Canada. People are very pleasantly surprised when they try it. What amazes them is these are very young whiskeys, but they show a great deal of maturity, far more than their ages indicate. We see that as dispelling the whole notion that age is the only important indicator of flavor.

🥃 **How so? Why is that different for your whiskey?**

The environment in which we're aging our whiskey. Everything we're doing is done in the same facility. It's not climate controlled, so we get these extreme swings of temperature and humidity that cause a great deal of interaction between the spirit and the wood. It's much more active than in a more temperate climate like Scotland, where they lose about 2 percent angels' share. Here we have hot humid summers and cold dry winters, which leads to more loss, but it creates better interaction.

And that's made even more dramatic by what happens inside the distillery. For example, in the morning, we come into the distillery and it's cold and dry. But by the end of the day, it's warm and steamy in there because the stills are running. That's causing the spirit to interact with the wood at a much more rapid rate. We believe that's what's giving the whiskey such nice flavors and character. We did not know it would happen. It was a happy accident.

**ABOVE:** *Still Waters Distillery's single malt whiskey, made from Canadian two-row malted barley.*

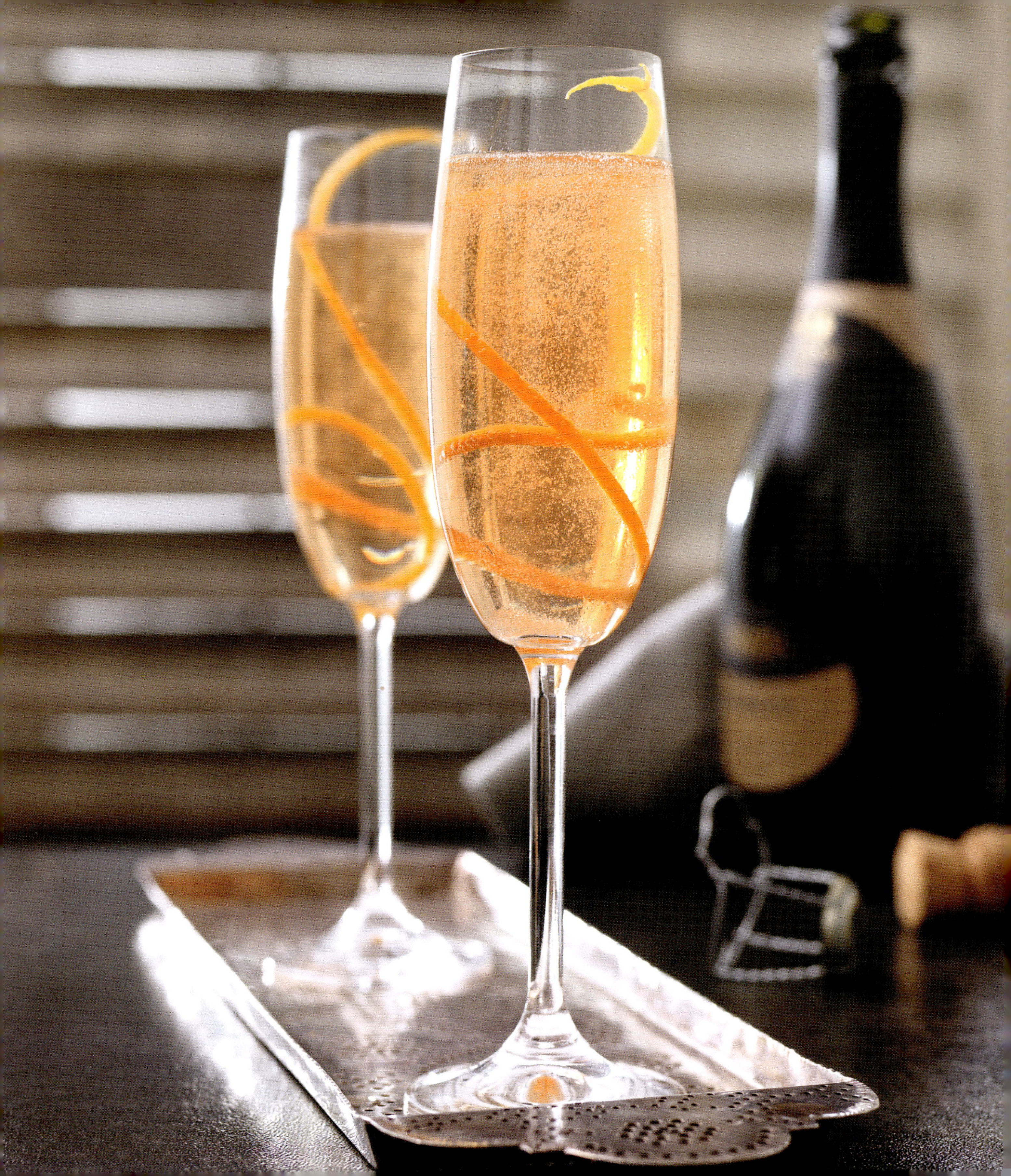

## THE SEELBACH

*This classy cocktail was created at the Seelbach Hotel in Louisville, Kentucky. It is not known whether F. Scott Fitzgerald was drinking these when he got loaded and was thrown out of its bar. What is certain is that it's possible to lose one's composure after a couple of these.*

INGREDIENTS

1 oz (30 ml) Old Forester bourbon, preferably Signature (100 proof)
½ oz (15 ml) Cointreau
4 dashes Peychaud's bitters

3 dashes Angostura bitters
Champagne or sparkling wine
Orange rind, to garnish

METHOD

Blend the bourbon, Cointreau and bitters in a chilled champagne flute, and top up with cold champagne or sparkling wine.

To garnish, peel a strip of orange rind using a vegetable peeler and twist over the drink to express its oils into the glass.

**LEFT:** *The Seelbach*

## BOURBON 'RITA

*In Kentucky it's our God-given right to drink bourbon, and there is a social expectation that we should do so. For those outsiders — such as my wife — who need a little help with this, I took my standard margarita recipe and replaced the tequila with bourbon to make a popular party drink.*

INGREDIENTS

2 oz (60 ml) bourbon (preferably 90 to 100 proof)
3 oz (90 ml) sour mix (fresh-made is best)
½ oz (15 ml) orange liqueur
2 lime wedges

METHOD

To a small shaker half-filled with ice, add the bourbon, sour mix, orange liqueur and the juice squeezed from one lime wedge; cover and shake vigorously for about 10 seconds. Almost fill a rocks glass with ice and squeeze over the juice from the reserved lime wedge. Strain the bourbon mixture into the glass.

# THE PRESBYTERIAN

*It's hard to settle on a single story behind the name of this drink — Scotsmen aren't big on ginger ale and Americans tend not to mix good Scotch with anything — but the fact is that this combo tastes quite nice, especially for those working up to bold Scotches.*

## INGREDIENTS
2 oz (60 ml) Scotch
   (or bourbon or rye whiskies)
Ginger ale
Club soda

## METHOD
Fill a Collins glass with ice and follow with the Scotch. Judge the remaining space in the glass and add equal amounts of ginger ale and soda. Stir gently to blend; take care not to over-agitate, or you will release all the fizz.

# MINT JULEP

*The Kentucky Derby made this drink famous, but there's no place that makes a worse mint julep than the home of that heralded horse race, Churchill Downs. Long story short: never buy one there. And if you follow this recipe, you'll never have to.*

## INGREDIENTS
2½ oz (75 ml) bourbon (at least 90 proof but
   preferably 100)
Leaves from 4–5 mint sprigs
2 sugar cubes or ½ oz (15 ml) simple syrup
   (1:1 water to sugar)
Mint sprig, to garnish

## METHOD
Place the mint leaves and sugar cubes (or simple syrup) into a julep cup or Collins glass; muddle gently to dissolve the sugar and to release the mint oils (muddling too firmly can make the mint somewhat bitter). Fill the glass with cracked ice (crushed ice melts too quickly) followed by the bourbon; stir well to begin dilution, but stop when the glass becomes frosted. Garnish with the mint sprig.

**RIGHT:** *Mint Julep*

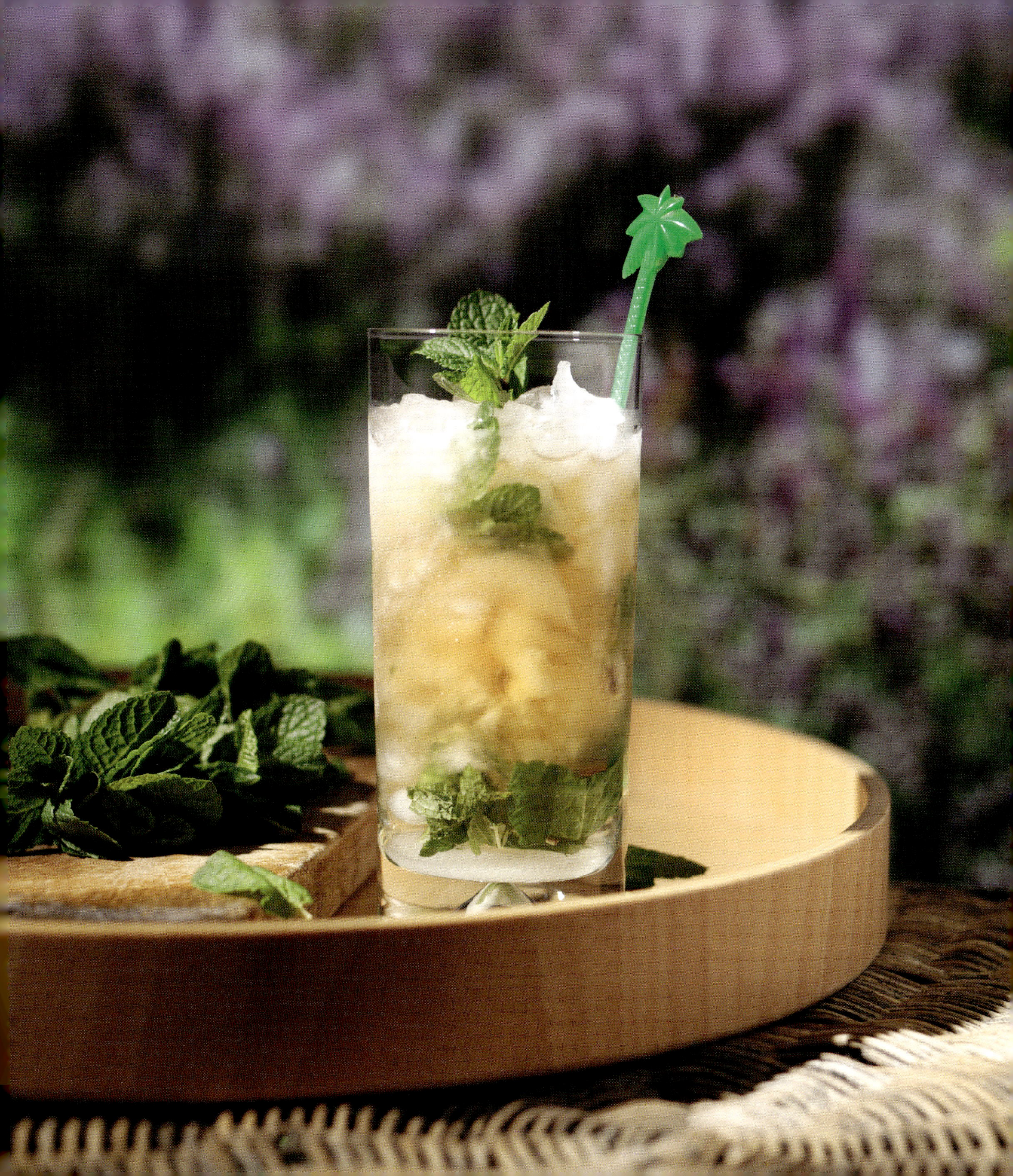

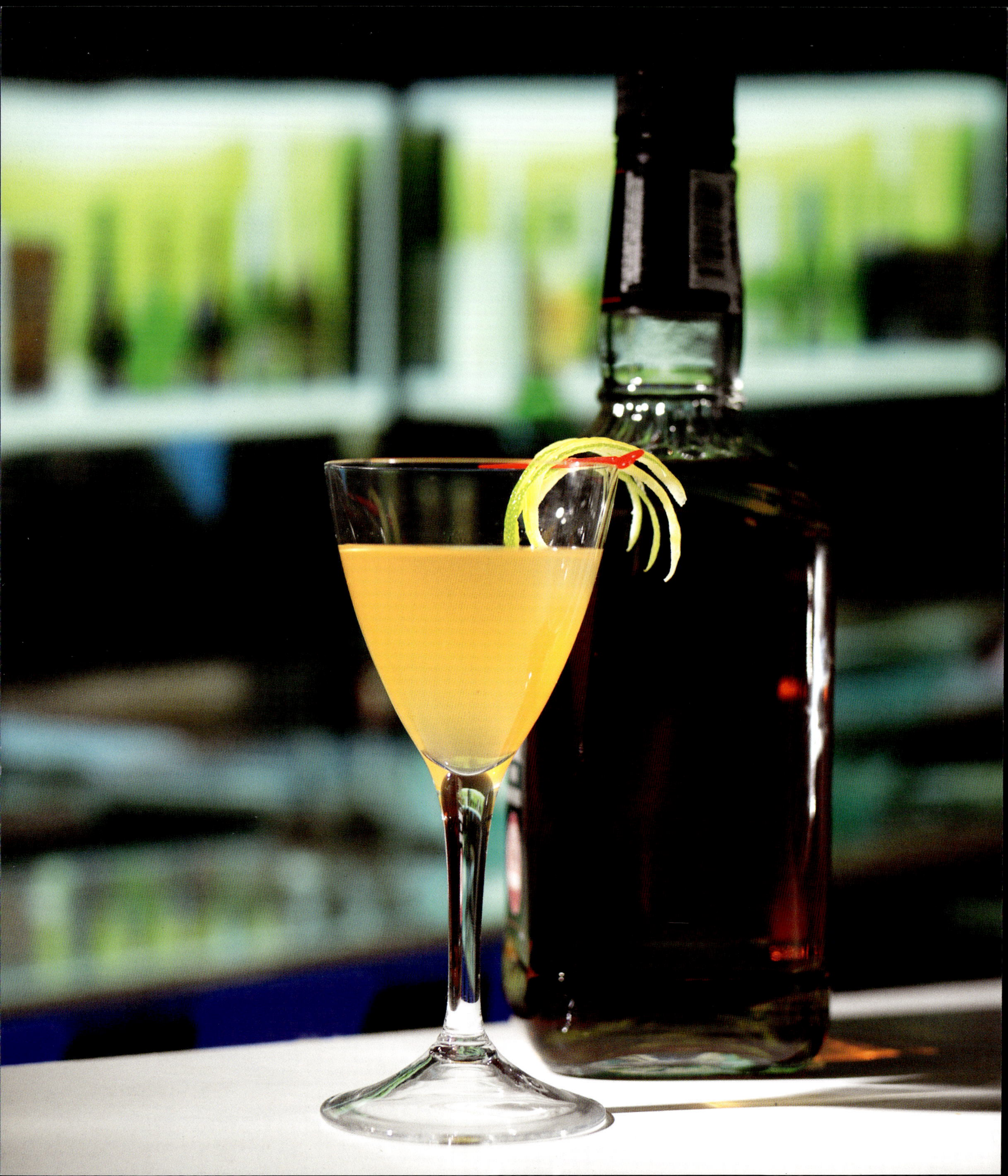

## WHISKEY SOUR

*So you think you know how to make a proper version of this old-timey cocktail? Are you using egg white or cheating with sour mix? Do it right with freshly-squeezed lemon juice, and for real depth, substitute sorghum molasses for sugar when making the syrup.*

### INGREDIENTS
2 oz (60 ml) whiskey
1½ oz (45 ml) freshly-squeezed lemon juice
1 oz (30 ml) simple syrup (1:1 water to sugar)
1 tbsp (15 ml) egg white
1 orange slice, to garnish

### METHOD
To a small shaker add the whiskey, egg white and simple syrup; cover and shake vigorously for about 5 seconds. Fill halfway with cracked ice, add the lemon juice, cover and shake again for 5 seconds to froth and incorporate. (Adding the lemon juice out of sequence can clot the egg white so don't be tempted to combine these steps.) Strain into a chilled rocks or coupe glass and garnish with a skewered orange slice.

**LEFT:** *Whiskey Sour*

## I'LL BLACK YOUR RYE

*Sometimes a good pun is the result of a good drink ... or three. Modern bartenders are not only working with fun ingredients, they're getting inventive with names too. With this is amalgam you can swap the rye whiskey for brandy, but the name won't work so well!*

### INGREDIENTS
1 oz (30 ml) rye whiskey
2 oz (60 ml) Malbec or Zinfandel white wine
½ oz (15 ml) lemon juice
½ oz (15 ml) agave syrup
   (1:1 water to agave nectar)
3 blackberries
Lemon rind, to garnish

### METHOD
To a small shaker add the blackberries, agave syrup and lemon juice, and muddle. Add several ice cubes followed by the rye whiskey and wine. Cover and shake vigorously for about 5 seconds, and then strain through a fine mesh strainer into a coupe glass.

To garnish, peel a strip of lemon rind using a vegetable peeler and twist over the drink to express its oils into the glass.

# RUM

When we hear the word "by-product," we commonly assume it's a reference to waste or even garbage. And for centuries, that's exactly how Caribbean sugar farmers regarded molasses. They knew how to crush cane, capture its cloudy juice and boil it down to shelf-stable crystals, and they also knew its liquid form could be distilled into rum. But as for the dark sticky liquid left behind from the refining process, they chucked it into the ocean or fed it to slaves and livestock (according to Wayne Curtis, author of *And a Bottle of Rum*). Thankfully, someone eventually figured out molasses could be distilled into alcohol, and suddenly that black sticky by-product became a highly profitable revenue stream. Today it remains one of the world's most popular spirits, serving as the core of many cocktails and punches, or when well-aged, sipped neat.

**RIGHT:** *12-year-old rum, made from fermented and distilled sugar cane by-products and aged in oak barrels.*

There's nought, no doubt, so much the spirit calms as rum and true religion.

— LORD BYRON

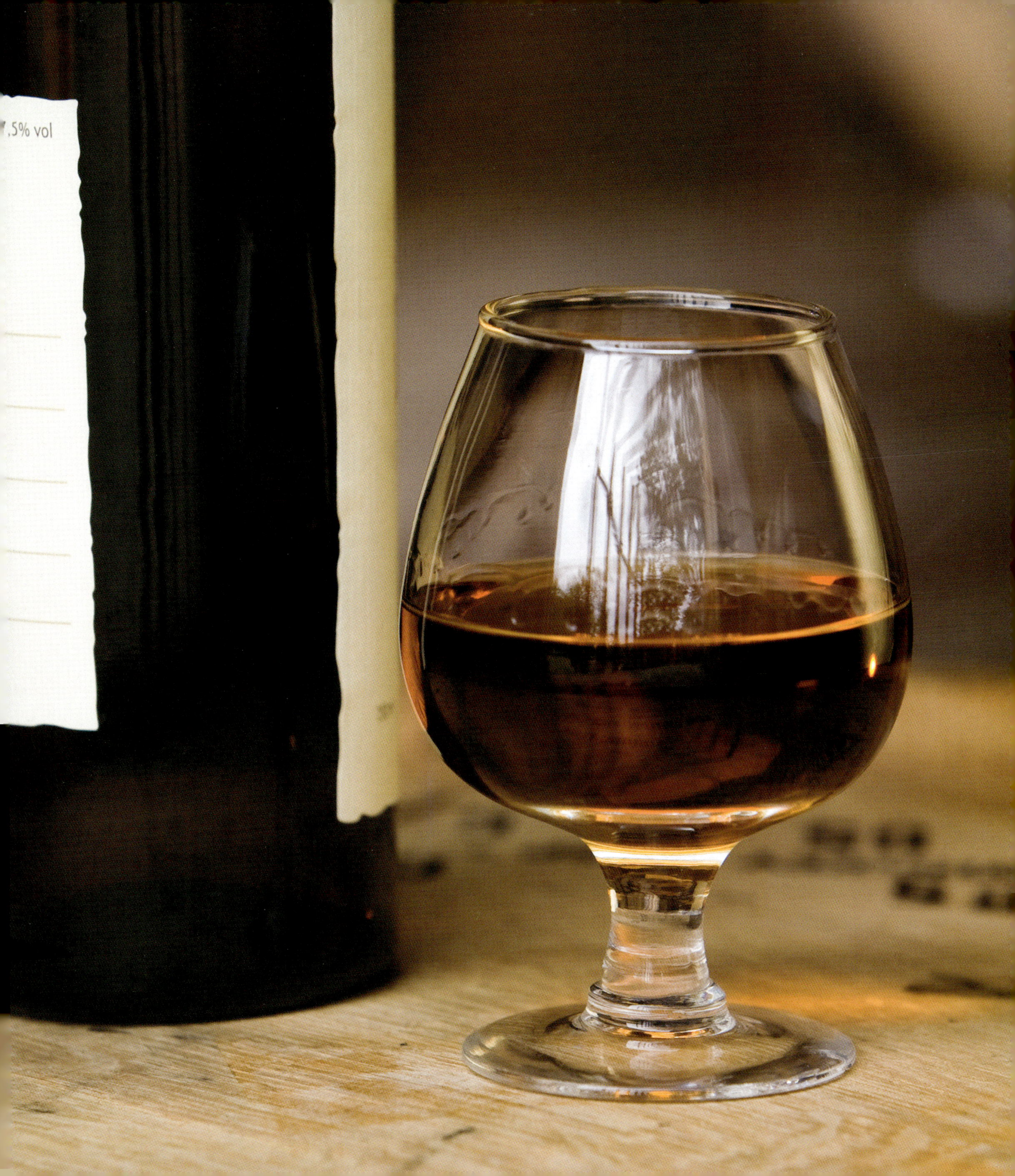
.5% vol

## TYPES OF RUM

In the 18th century, the navies of Spain, France and England coursed the Caribbean regularly, and so the trade of rum surged. While at work navigating the turquoise waters, sailors in the English Royal Navy were rewarded with a daily ration of rum that ranged from one to two pints, and British colonists in America were such fans of rum that they began importing molasses and distilling their own.

Rum is widely recognized as a product of the Caribbean island nations, where sugarcane grew in abundance, yet its origins are easily traced to the Far East; to China, India and even Iran. And while sugarcane is still grown throughout the Caribbean, a sizable portion of the world's rum is produced from molasses exported from Brazil, Thailand and India.

Rum can be made anywhere, and a good deal of it is produced in Canada. But unlike Caribbean rum, which is made strictly from molasses or crushed cane juice, Canadian-made rums come from blends of molasses-based neutral spirits and imported West Indian rums.

There are four basic types of rum made from molasses, and then there is rhum agricole, a rum made from the juice of freshly-crushed sugarcane.

### White (Light) Rum
These rums are not aged, and they're bottled at proofs ranging from 80 to a fiery 140. Sipped straight, white rum can burn and deliver a cane-sugary note that's unusual to palates conditioned to refined sugar sweetness. But when added to a cocktail or rum punch, they're fantastic, giving juices and mixers edge and backbone that no other spirit can replicate.

### Golden Rum
Darker rums don't commonly carry age statements on their labels, so there is no standard for these. Sure, one could assume that the golden color would signal a rest somewhere between three and nine months in wood barrels. But since caramel coloring also can be added to aged rums, such assumptions are risky. The good news is golden rums are softer than their white counterparts; they, too, are delightful in cocktails and punches, and some are even decent sippers.

### Dark Rum
Again, one assumes that the deep amber of dark rum signals a Rip Van Winkle-like rest in bourbon, cognac or whiskey barrels — and sometimes that's true. But the use of caramel coloring and other flavoring to enhance these products makes any age assumption suspect. What is safe to assume, however, is these are great consumed neat in a Glencairn glass or snifter, or in a rocks glass with a cube or two of ice. Despite the use of additives, these spirits showcase rum distillers' best work and reflect the vigorous barrel activity effected by the Caribbean's hot weather.

### Spiced Rum
These typically are dark rums with added spices. Some commonly used flavors are: allspice, clove, nutmeg, cinnamon, orange peel and vanilla.

### Rhum Agricole
While it can be made anywhere, rhum agricole on the island of Martinique is made in three ways: blanc, aged three months; elevé sous bois, aged 12 months; and vieux, aged at least three years. All of these are delicious and delicate rums, regardless of age.

> **A WORD ON COOPERAGE**
> Rum distillers most commonly choose used bourbon barrels for aging since new charred oak barrels impart so much flavor that they dominate the delicate sugar spirit. Lately, though, some distillers are experimenting with new barrels and to good effect. Other cooperage includes barrels first used for American whiskey, sherry or cognac.

## FERMENTATION

Since fresh cane juice isn't always readily available, many rum washes begin with a blend of molasses and water. Some recipes call for equal amounts of molasses and white sugar added to water. Either way, you'll have plenty of sugar to ferment.

**ABOVE:** Sugar cane is crushed for its juice in Grenada.

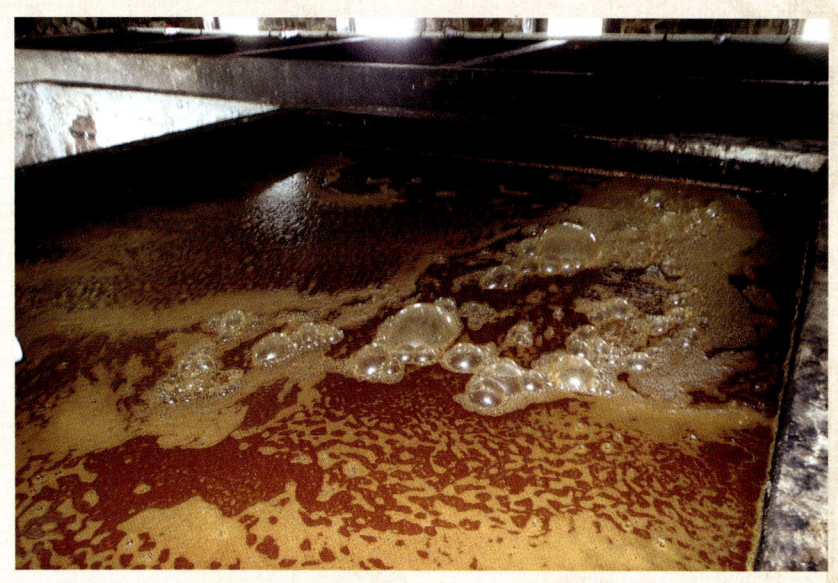

**ABOVE:** Sugar cane juice bubbles as it ferments in an open fermenter at Rivers Run Distillery, Grenada.

Once blended, the wash is heated to 170°F (76°C) and then cooled to 80°F (28°C) to safely accommodate and activate the yeast. In Word's of Wisdom (pages 88–89) with Heather Bean, owner of Syntax Spirits Distillery, she talks about how she discovered that while molasses is full of sugar, much of that is maltose, a long-chain unfermentable sugar that's not readily accessible. The experienced whiskey maker and chemical engineer deduced that adding glucoamylase enzymes might free those sugars, and she was right. In her words, "It worked brilliantly." So consider this if you are fermenting an all-molasses rum.

Once your wash is fermented, use your hydrometer to determine your potential alcohol and prepare to move to the still.

## DISTILLATION

Depending on the brand, commercial rum is distilled in pot stills (often with doublers) and column stills or both. Their finished alcohol percentage ranges between 70 percent and 95 percent. Not surprisingly, many fine vodkas are distilled from a base spirit of rum. Many rum fans also swear that the best are made in pot stills, which give the spirit a bit of oblique character and discernible mouthfeel.

As with nearly all ferments, it is important to schedule your time to do a stripping run and a finishing run.

### The Stripping Run

Siphon off the fermented liquid from your wash and into your still's kettle. Reassemble the still and gradually begin heating the wash to start your stripping run. Heating it slowly will reduce the chance of scorching any residual solids carried over from your fermenter, and also lessens the wear and tear delivered from high, direct heat coming from the base of the still.

Shortly afterward, start running cold water into your condenser. This is a foolproof way of ensuring your alcohol will be condensed rather than vented into the air as flammable vapor and lost alcohol. If you choose not to start your water so early, at least have a checklist to follow, or set a timer to remind you to turn

**ABOVE:** *Distillation at Rivers Run begins with a pot still (far right), then moved to a pair of doublers before and then condensed.*

it on. Experienced distillers will advise you to never leave the room once your still is started. Your initial target temperature will be 174°F (79°C), which is when methanol begins to boil. Once reached, allow the temperature to rise steadily to a maximum of 207°F (97°C), when the tails cut finishes boiling off. In order to avoid distilling the remaining water in your wash, make sure you do not exceed this temperature.

Since this is just a stripping run, where you separate the alcohol from the water in the wash to make a low wine, it won't be necessary to make cuts. Collect all the alcohol from this run, cool down your still, disassemble it, then clean it, before reassembling it and adding the alcohol back to it to begin your finishing run.

### The Finishing Run

Before you begin your finishing run, take the time to label your collection vessels, "Heads," "Hearts" and "Tails," to ensure you do not mix up your cuts. Then gradually start the still and heat the wash.

As your wash heats up and the still begins to run, collect and discard your foreshots and heads. Most distillers will advise collecting the first 4 ounces (120 ml) of distillate from a 5 gallon (19 l) wash and discarding it. Obviously, the amount of heads and foreshots increases or decreases based on the size of the wash, so make sure you do the math ahead of time. Hillbilly Stills president Mike Haney advises discarding the first 2 percent of liquid emerging from the still (see page 66).

**ABOVE:** *Still and doubler in a commercial rum distillery.*

the condensed distillate, it accumulates just enough to float the alcohol hydrometer. The constant flow of spirits allows the distiller to determine regular proof readings as soon as they leave the still. The distillate will then overflow out of the parrot and into the final collection vessel.

Continue collecting hearts until the alcohol drops to about 65 percent. Set that aside and begin collecting tails. Distillers do this for two reasons: to use them for flavoring the final distillate; and to redistill later to extract any residual ethanol.

### Blending and Storing

If you want to blend some of your tails into your hearts, place all your hearts in a single container and stir in small amounts of tails to suit your personal tastes.

Once finished, "proof down" your rum by diluting it with distilled water and transferring it into glass or stainless storage containers or, if you're aging it, to your barrel. If you're not aging your rum, then proceed directly to the refrigerator and create a delightful rum punch.

As the heads continue to come off the still, check the proof with an alcohol hydrometer. Ideally, the proof should be about 80 percent alcohol. When that proof begins to drop, you will be distilling hearts. Fill your graduated cylinder halfway and float your alcohol hydrometer to watch the proof drop as the still continues to run. Keep regularly tasting the distillate to discover the flavor of the spirits at each point by making cuts with your nose and tongue. Most distillers agree that this is the most reliable way of determining flavor.

You also can then eliminate the handheld alcohol hydrometer by using a still parrot; a fixed cylindrical pipe and collection tube welded between the condenser and your final collection vessel. As the parrot collects

**ABOVE:** *Syntax Spirits' prized white rum (see page 89).*

# CLASSIC DAIQUIRI

-------------

*Due to the advent of the blender, this cocktail has suffered more abuse at the hands of barkeeps than the margarita — avoid its frozen form. Named after a beach on the island of Cuba, it was a favorite of literary boozehounds Ernest Hemingway and F. Scott Fitzgerald.*

## INGREDIENTS

1½ oz (45 ml) light rum
¾ oz (22 ml) freshly-squeezed lime juice
¼ oz (7 ml) simple syrup (1:1 sugar and water)
1 lime wedge, to garnish

## METHOD

Half-fill a small shaker with ice and add all the liquid ingredients. Cover and shake vigorously for about 5 seconds. Fill a rocks glass halfway with ice and strain the cocktail into the glass.

Garnish with the lime wedge.

# HEMINGWAY DAIQUIRI

-------------

*Named after Ernest Hemingway, one would find it predictable that "Papa" would prefer a high liquor-to-mixer ratio, but the addition of a little maraschino liqueur or grenadine seems surprisingly soft for such a rough-and-tumble sort.*

## INGREDIENTS

3 oz (90 ml) light rum
1 oz (30 ml) freshly-squeezed lime juice (about half a juicy lime)
½ oz (15 ml) grapefruit juice
¼ oz (7 ml) maraschino liqueur or grenadine

## METHOD

Half-fill a small shaker with cracked ice and add all the ingredients. Cover and shake gently for about 5 seconds.

Strain into a chilled rocks glass.

**RIGHT:** *Hemingway Daiquiri*

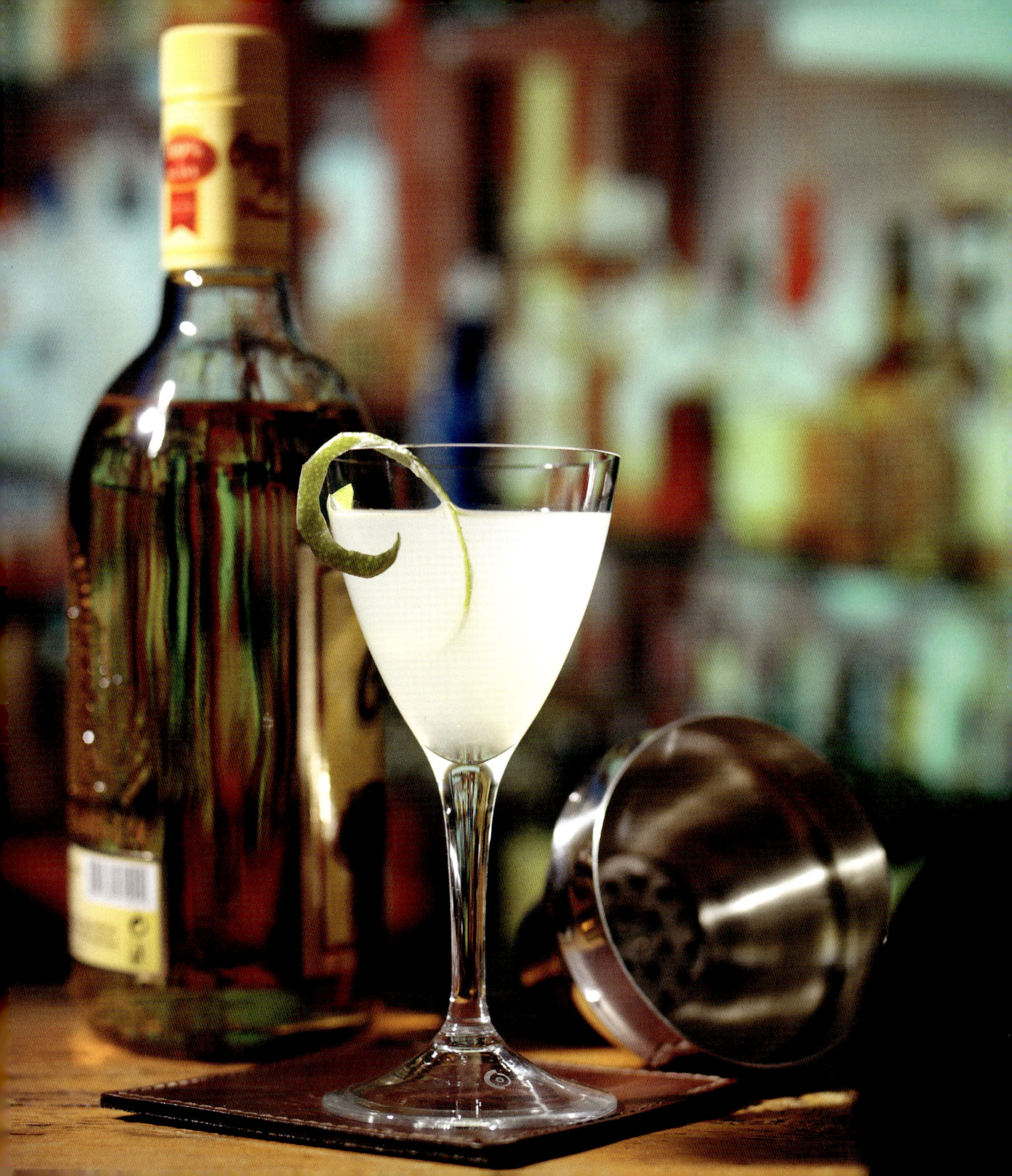

# WORDS OF WISDOM

## HEATHER BEAN
### MISTRESS OF STILLS, SYNTAX SPIRITS DISTILLERY
Greeley, Colorado
www.syntaxspirits.com

Heather Bean's "rumored" home still didn't challenge her chemical engineer's mind. So she launched her own distillery in 2011.

**Did you ever hobby distill at home?**
I may or may not have had a really horrible pot still on my stove at some time in the distant past. And I won't exactly deny playing around with a spool of copper tubing or anything like that; maybe that's possible. The real story is: I'm a chemical engineer and I wasn't satisfied with anything I could do with spirits at home. I only have very low ceilings at home, which is a problem.

**So you took the plunge and opened your own legal distillery.**
Yeah, I figured I would go big or go home. Well, not Budweiser big, but big enough to do some volume. I was brewing my own beer, but the chemical engineer in me wanted to distill. I had to try distilling.

There are two stories about why I got into this professionally. The real story is home distilling was wildly impractical for what I wanted to do; I understood scale and knew what was possible with a real distillery. But a more interesting story I've told is that working in a corporate environment had driven me to drink, and since what I liked to drink was getting too expensive, I turned to making my own. That was funny when it was a joke, but it kind of got out of hand over time when I made a career out of this. The truth is you can't get into this halfway.

**What's inspiring about your craft these days?**
Barrel aging is fascinating. It was something I couldn't appreciate until we started doing it three years ago. It's always a challenge and so much of it is out of your control. You'll have a whiskey that's going along great and then it changes completely. And you're thinking: oh no, what's wrong? And then it changes back again, and you're asking yourself why it does that. Distilling is a

mix of science and judgment calls. It's always very interesting.

🍶 **What's the biggest misconception hobbyists have about distilling?**

That you can put something crappy into the still and get nice alcohol out of it. Garbage in. Garbage out. Everything at every stage matters when you put it in the still: the yeast, the mash, all of it.

🍶 **You make vodka and bourbon, but let's talk about your rum, White Powder.**

We use all dark molasses from Florida — the one ingredient we can't get locally. Molasses is what's left over after they take out the white sugar — pure sucrose — which I don't think has any flavor anyway. Molasses contains a lot of maltose and other long-chain unfermentable sugars. But they lend the rum toasty and grainy notes, and that's what we wanted. We experimented with other sugars and cane juice, but we felt the flavors available to work with were pretty slim. I think some white sugar rum tastes like jet fuel, so we kept coming back to darker and darker molasses for flavor.

But when we first tried fermenting molasses, we discovered how little alcohol you get out of it despite its specific gravity. You see all that sugar, but you learn that a lot of it is unfermentable. But we figured out to use glucoamylase, an enzyme, to break apart those sugar chains. It worked brilliantly.

I also think that some of those really good flavors are the product of using column stills, where we get really purified spirits. I'd think that using it in a whiskey still could be too overwhelming, but that's a personal preference thing. I just happen to like distillate that explains a little about where those flavors you're tasting actually come from.

🍶 **Other distillers tell me rum-making can be challenging. Why do you think that is?**

For one, it takes longer to get it fermented correctly and completely. We also had to find the right yeast for it. Just like when we created our vodka, we tried

**ABOVE:** *Syntax's award-winning small-batch vodka and rum, distilled at their Colorado-based distillery.*

several specific types of yeast for flavor. But oddly enough, we wound up using a pretty generic yeast — the yeast we chose for our vodka.

🍶 **Are you aging some of your rum?**

Yes, we are, some in wine barrels and some in bourbon barrels. It seems to be going faster than the whiskey we're aging, and maybe that's because it's a much lighter flavored spirit than whiskey. That's also why we like used barrels for aging. I suspect a new barrel could completely overwhelm the rum, which is more delicate. Our rum has been in barrels for about a year, and we think we'll be bottling soon so it doesn't overage.

## CABLE CAR

*Created in 1996 by famed bartender Tony Abou-Ganim, the recipe for this drink most often calls for Captain Morgan spiced rum. There are a growing number of spiced liquors now available that would make a superb substitute, such as Corsair Distillery's gin on page 106.*

INGREDIENTS
1½ oz (45 ml) Captain Morgan spiced rum
¾ oz (22 ml) orange curacao
1½ oz (45 ml) sour mix (fresh-made is best)
1 orange wedge
Coarse sugar
Orange rind, to garnish

METHOD
To sugar rim the glass, take a small wedge of orange and rub it around the rim of a chilled rocks glass. Pour coarse sugar into a small bowl and dip the rim of the glass in, moving it around to coat the edges evenly both inside and out. Fill the glass about halfway with ice, taking care to avoid touching the sugared rim.

Half-fill a small shaker with ice, add the liquid ingredients and shake vigorously for about 5 seconds. Strain the drink into the prepared glass.

To garnish, peel a strip of orange rind using a vegetable peeler. Fold to express its oils into the glass.

## DARK AND STORMY

*Perfect summer cocktails are easy with this poolside sipper from Bermuda. Gosling's Black Seal is the traditional rum for this, but any good dark rum will suffice, and Barrit's, a Bermudan ginger beer, is great if you can find it. The addition of lime juice adds a zesty touch.*

INGREDIENTS
2 oz (60 ml) Gosling's Black Seal rum
3 oz (90 ml) ginger beer
1 lime wedge

METHOD
Fill a rocks glass with ice, add rum and a squeeze of lime and stir for 10 seconds with a bar spoon. Add ginger beer while continuing to stir gently to blend. Alternatively, to produce a nice dark and light contrast in the glass, don't blend in the beer — after a few sips it'll blend itself.

**RIGHT:** *Dark and Stormy*

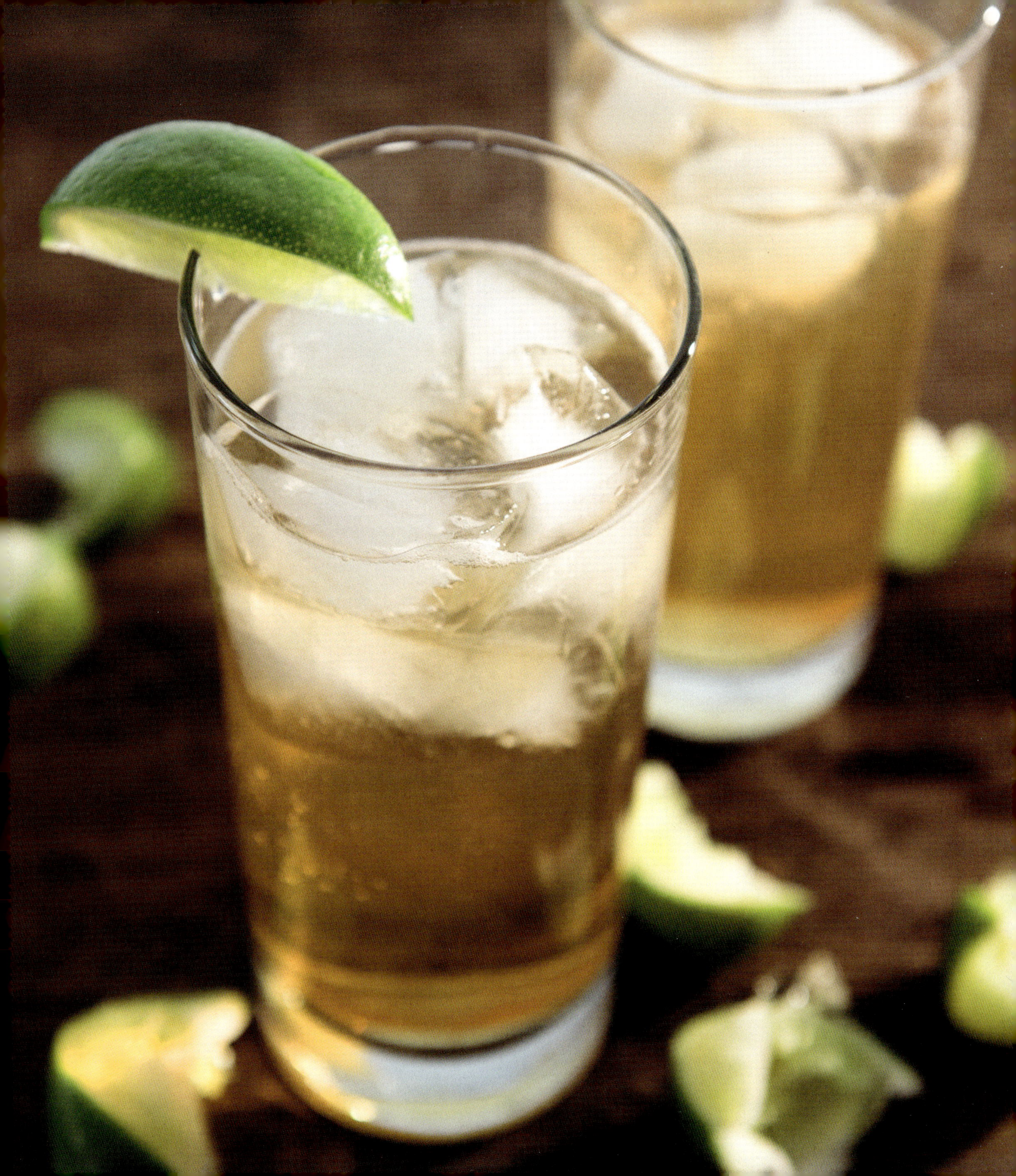

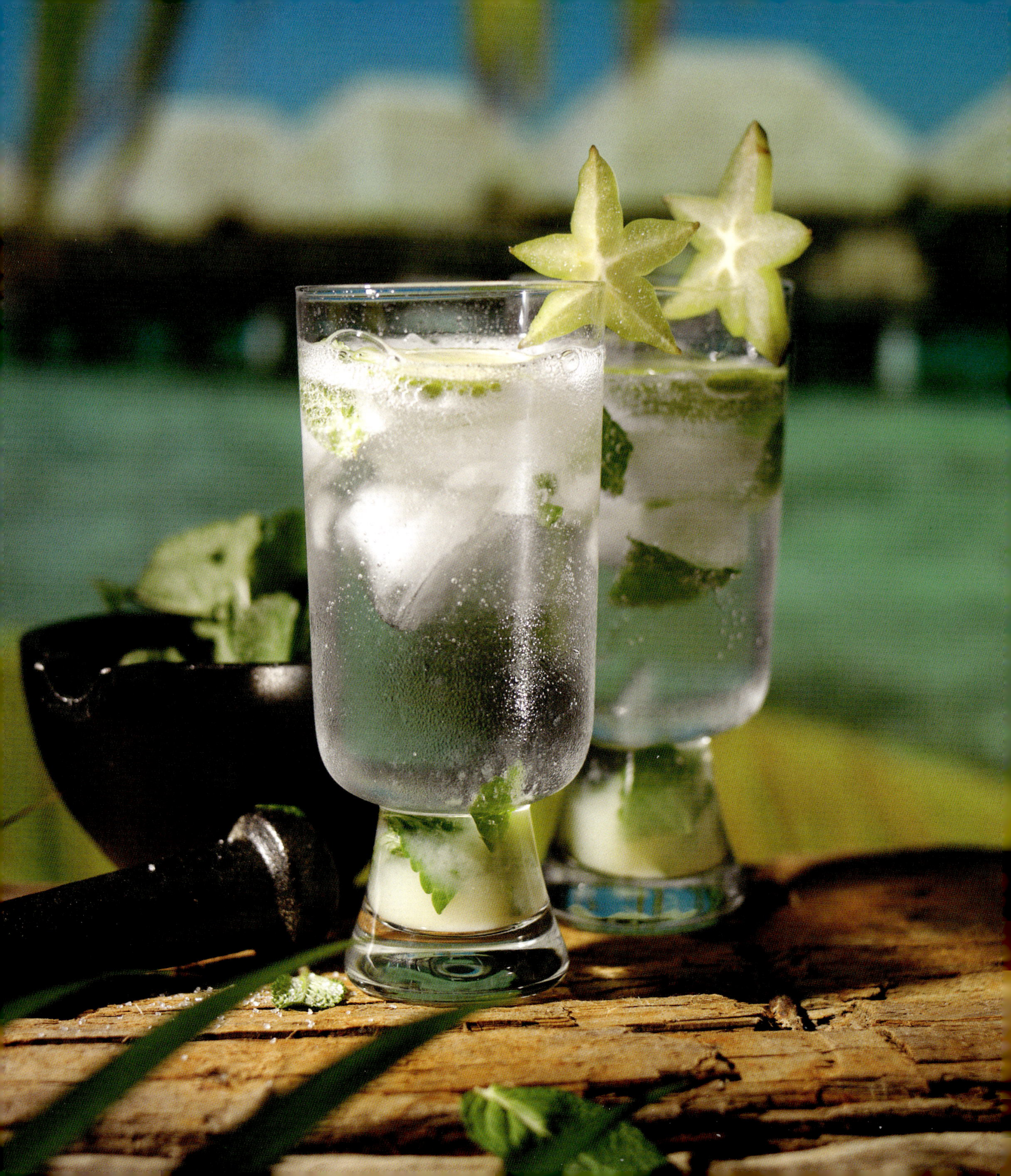

# MOJITO

*There's no shortage of varied stories attached to this simple cocktail's origin, but what they all agree on is that it's a Cuban creation. It gets its sweetness from freshly-crushed sugarcane juice called* guarapo, *but a simple syrup made with granulated sugar works just as well.*

INGREDIENTS

2 oz (60 ml) light rum
1 tbsp (15 ml) simple syrup
   (1:1 water to granulated sugar)
6–8 mint leaves
Club soda
1 lime
Mint sprig, to garnish

METHOD

Add the mint leaves and simple syrup to a Collins glass and muddle gently but thoroughly, bruising the leaves without tearing them. Use a juice press to squeeze the juice of two halves of a lime into the glass, discarding the first and dropping the second into the glass.

Fill the glass with ice, add the rum and stir for 10 seconds with a bar spoon. Top off with club soda, stirring gently to blend. Garnish with a mint sprig.

**LEFT:** *Mojito*

# HURRICANE

*Pat O'Brien's bar in New Orleans is credited with creating this concoction from surplus rum that was forced onto bartenders during World War II. To buy a ration of short-supply Scotch, they had to take cases of rum too — Pat O's figured out just how to use that surplus.*

INGREDIENTS

2 oz (60 ml) light rum
2 oz (60 ml) dark rum
2 oz (60 ml) passion fruit juice
2 oz (60 ml) sour mix
1 oz (30 ml) orange juice
1 oz (30 ml) freshly-squeezed lime juice
   (about half a juicy lime)
1 tbsp (15 ml) simple syrup (1:1 sugar and water)
1 tbsp (15 ml) grenadine
1 cherry, to garnish
1 orange slice, to garnish

METHOD

Half-fill a small shaker with cracked ice, add all the liquid ingredients and cover and shake for about 5 seconds. Strain into a hurricane glass (if you have one) or a Collins glass half-filled with ice. Garnish with a cherry and an orange slice.

# COMMODORE

*Just as some say you should never have to specify "gin" in a martini, you'd expect Commodore to always include rum (light or dark), but bourbon, rye or Canadian whiskey may often substitute. Here I am using rum, of course.*

INGREDIENTS

2 oz (60 ml) light or dark rum
1 egg white
½ tsp sugar
Squeeze of lemon juice
1 dash of grenadine

METHOD

Half-fill a small shaker with cracked ice and add all the ingredients. Cover and shake vigorously for about 20 seconds to fully froth the egg white and strain into a chilled glass.

# BOLO

*This speakeasy Prohibition favorite is one of many nearly forgotten sippers now making a comeback. The key to a good one is (like so many) freshly-squeezed fruit juice. I add both lime and orange juice to give mine a good flavor.*

INGREDIENTS

2 oz (60 ml) light rum
½ oz (15 ml) freshly-squeezed lime juice
   (about a quarter of a juicy lime)
1 oz (30 ml) freshly-squeezed orange juice
   (about half a juicy orange)
2 dashes Angostura bitters

METHOD

Using a juice press, squeeze the lime and orange juices into a small shaker half-filled with ice. Add the rum and bitters, cover and shake vigorously for about 5 seconds. Strain into a chilled rocks glass.

**RIGHT:** *Bolo*

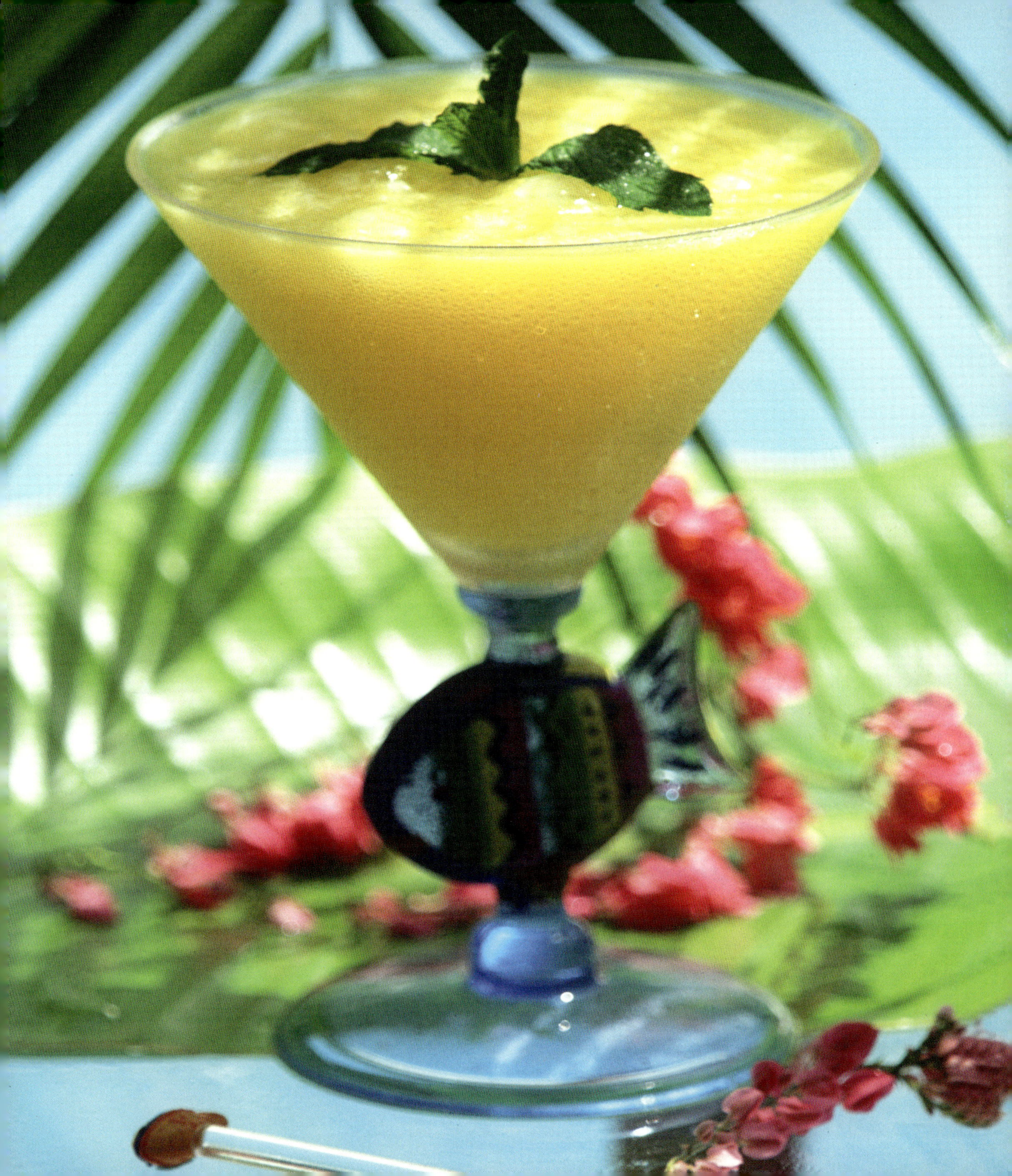

# HOT BUTTERED RUM

*Arguably a mulled cocktail, this centuries-old simple blend of rum, butter, hot water, sugar and spices seems a divinely inspired way to ward off a cold in winter. For extra punch, you can add a splash of cider with the hot water.*

### INGREDIENTS
2 oz (60 ml) dark rum
1 tsp softened butter
1 tsp brown sugar
1 tsp (5 ml) vanilla extract
Spices to taste: ground cinnamon, nutmeg and allspice
Hot water

### METHOD
Add the butter, sugar and spices to an Irish coffee glass or a heavy walled mug and muddle well. Add rum and hot water and stir to incorporate.

# CAIPIRINHA

*Caipirinha is the national drink of Brazil and the word translates roughly as "bumpkin," a warning to its potency, no doubt. Just as the Hurricane is ubiquitous at Mardi Gras, this is a standard at Carnavale. It is made with cachaça, a super-sweet Brazilian rum made from crushed sugarcane juice.*

### INGREDIENTS
2 oz (60 ml) cachaça
1 lime, quartered
2 tsp fine sugar

### METHOD
Put the lime wedges and sugar into a rocks glass and muddle well. Fill the glass with ice cubes, add cachaça and stir well for 15 seconds using a bar spoon.

# GIN

------

*It's a fact that drinkers either love or hate gin, and with equal passion. Brown spirits fans fonder of creamy textures and caramel notes wrought from a long rest in the barrel often don't appreciate the botanical crispness of gin and its in-your-face bracing nature. Dislike of this lovely spirit sees them miss out on the pleasure of a classic martini or arguably one of the best cocktail pairings ever, gin, tonic and lime. And why? Blame it all on juniper, gin's essential and prominent flavoring ingredient. The diminutive evergreen conifer packs the bright herbaceous character for which gin is revered and reviled, allowing it to cut through and stand tall on its own or in a cocktail. Gin drinkers accept this somewhat abrasive character as proof of their love for savory sippers over sweet ones.*

**RIGHT:** Classic gin and tonic, served with lemon and ice.

I exercise strong self-control. I never drink anything stronger than gin before breakfast.

— W.C. FIELDS

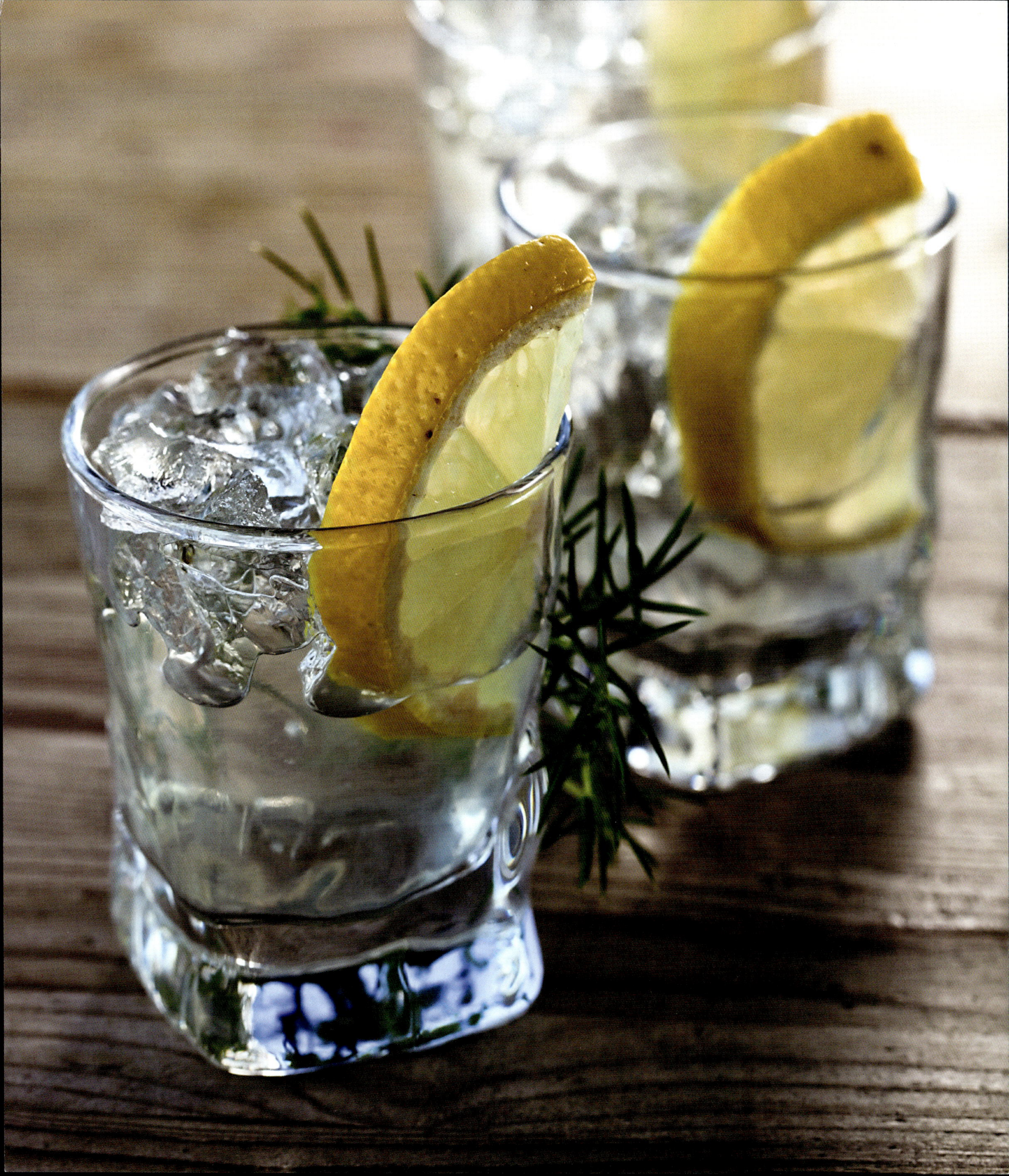

## A LITTLE HISTORY

The origins of modern gin can be traced back to genever, a drink created in Holland in the 1500s, with production spreading to Belgium in the 1600s. The word genever is roughly translated as juniper, but in its earliest forms it bore little resemblance to today's gin, which is so crisp on the palate.

Since brandy was already popular throughout Europe in the 1500s, some genever was made from grapes. But that eventually changed when distillers began fermenting grains such as rye, barley and corn in the 1600s. (Fast forward a couple of centuries, and some gin makers were fermenting and distilling beet sugar to make genever, but as drinkers and distillers agreed that the product was inferior to grain-based distillates, the practice largely ceased.)

Interestingly, genever bore more commonalities with whiskey than modern gin. Though its prominent note was juniper, its texture was softer and its taste was slightly malty. Overall, it was a gentle spirit to drink. It wasn't until the 1700s that distillers increased the load of botanicals the spirit became more complex and leaner bodied, as we know it today.

While genever was very popular in Holland and Belgium, no country embraced it quite so powerfully as England. When distillers there began producing it in large quantities in the 17th century, sales of it would soon surpass beer. The name also would be shorted to "gin," and distillers began to monkey with the mash bill too. By the early 18th century, the English were consuming an astonishing amount of the stuff. In 1733 alone, London distillers produced 11 million gallons of legal gin, which, according to Lesley Jacobs Solmonson, author of *Gin: A Global History*, amounted to 14 gallons (53 l) per person per year. The Gin Craze, a multi-decade booze binge, was a decidedly low point in that nation's history.

To make matters worse, most of the gin produced in England then was of poor quality. Some distillers bypassed traditional Dutch techniques by stretching their distillates with toxic substances such as turpentine and sulfuric acid and alum. No surprise that such poor quality gin came to be known as "rotgut." It was also common to sell gin at 160 proof, achieving the goal of intoxicating drinkers rather than dazzling their palates with dynamic flavors.

England's leaders believed they could address the overconsumption problem — and the country's revenue shortfall — by raising taxes on gin production and sales. The Gin Act of 1736 introduced steep licensing fees for retailing, and created fines for distilling gin at home. Subsequent laws rewarded those who reported illegal home distilling to authorities.

The Gin Craze did level off near the middle of the 18th century, but production continued climbing as exports of the spirit increased. As England's empire expanded, so did the distribution of gin, and in pre-Prohibition America it was a cocktail favorite.

Its post-Prohibition popularity was another matter. Like so many quality spirits made in the U.S., many disappeared during the 13 year legal distilling shutdown. Recovery of gin consumption was slow and noticeably hampered in the 1960s and 1970s with the surging popularity of vodka, but the 1980s was a turning point for spirits generally. With increased international travel, drinkers were exposed to better-crafted products that they wanted from restaurants, bars and liquor stores back home. Not only did gin exports increase as a result, small-batch distilleries eventually sprung up and gave life anew to the historic spirit.

## TYPES OF GIN

Unlike whiskey, whose preparation has remained largely the same after many centuries, gin has evolved profoundly since its creation. First came genever, followed by the lighter Old Tom, London dry and Plymouth gin, and finally the latest iteration of gin, New Western-styles, entered the juniper fray in the 1990s. What's remarkable is all five styles are still produced today, and the oldest among them are regaining popularity. Let's take a closer look at each.

### Genever

Genever is distilled from a malted grain mash bill, similar to whiskey. It's commonly bottled at 70 proof to 80 proof and aged in oak casks for one to three years. Though clearly juniper based, its grain influence

**ABOVE:** *Distilling equipment used in gin production is very similar to that used in the distillation of vodka.*

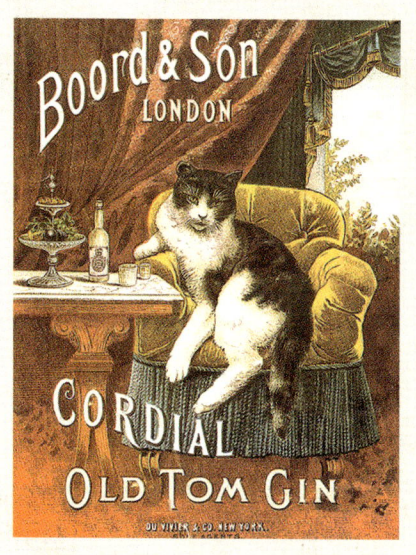

**ABOVE:** *Old Tom Gin poster, Boord & Son Distillery, London.*

is readily available to the palate through its medium body and malty, slightly sweet flavor. This makes it an exceptionally versatile cocktail liquor.

## Old Tom

Correctly regarded as the bridge between genever and London dry gin, Old Tom created a soft and inviting path from the old to the contemporary. Sweeter than genever (sometimes due to added sugar, other times from the use of sweeter botanicals) but more heavily influenced by botanicals, it was an easy sipper at 50 proof when created in the 1700s. But its punch grew more powerful over the next century, rising to between 80 proof and 90 proof. Its inherent sweetness made it an easy choice as a base spirit in many sweet cocktails created around the turn of the 20th century. Its return to the bar has been heartily welcomed by bartenders everywhere.

## London Dry

This stronger (80 proof to 114 proof), aromatic and unsweetened spirit defines gin today. London dry is

## HOW OLD TOM CAME TO BE NAMED

In the window of a London public house in the 1730s hung a sign with the image of a black cat painted on it. Extending from below the cat's paw was a pipe, and nearby was a slot into which passersby could place a coin, receiving in return a small drink of gin. The idea spread, and many public houses began to feature the tomcat's image outside their doors. When a drinker drew near, he would say, "Puss." And if someone on the inside replied, "Mew," the drinker could buy bootleg gin inside. The cat branding, subtle as it was, then led to gin being nicknamed Old Tom.

versatile and can be served on its own, star in a martini or take a supporting role in a cocktail. Its botanical collection always includes the dominant juniper followed by, but not limited to: coriander, angelica, orange peel, lemon peel, cardamom, cinnamon, cubeb berries, licorice, grains of paradise, nutmeg, and much more. Bombay Sapphire, credited by many as pivotal in gin's modern explosion, features its botanical lineup printed on each bottle. Some of the great London dry producers still operating today were founded in the 1800s, including Tanqueray, Beefeater, Gilby's and Gordon's.

### Plymouth Gin

This lower-alcohol gin (82 proof) is, as the name makes clear, produced in the port of Plymouth, located on the English Channel. Highly aromatic, it is slightly fruity and finishes softer than London dry. Like Old Tom, it's a great choice for introducing gin to those who've not yet caught on.

### New Western Gin

Pushing the boundaries of what has defined gin for centuries, New Western gins are still anchored by juniper, but other botanicals often get equal billing in the nose and on the palate. In some cases, like Corsair and Death's Door, the result is a spicier character. In others, like Hendrick's, Aviation and Martin Miller's, defining flavor notes are clearly vegetal and often represented by cucumber. While pine is fine with such distillers, juniper clearly is just one botanical on the team, no longer its lone star.

## DISTILLATION

It's a common assumption among home distillers that since gin begins with a grain-neutral spirit (GNS), it's simple to make. Professional distillers say otherwise.

"Coming up with a botanical mash bill is one of the hardest things in distilling," says Colin Blake, creative director at Distilled Spirits Epicenter (DSE), a distilling consulting and education firm in Louisville, Kentucky. In its gin classes, DSE displays 20 jars of botanicals commonly used in gin, but Blake says he has 34 total in inventory. "It's a pretty hard development process to get something that's good and right and balanced. A lot of distillery consultants won't even do gin."

But as the explosion of New Western gins is proving, new distillers can't resist the temptation to put their own spin on the spirit. Blake adds, "There are so many cool botanicals out there and so many ways to use them. So you've got all these craft guys experimenting on their own to figure out a recipe. But make no mistake, that's time-consuming and hard."

This is especially true if you insist on distilling your own GNS, which requires a vodka still and the careful selection of flavoring grains. "Even if you've got the equipment to make vodka, the effort required to get a few gallons will cost so much and take so much time they have to ask whether it's worth it," Blake says.

To achieve the spirit purity required for good gin, it'll require running the vodka through the still at least a few times, and that means a greatly reduced final yield. Blake advises, "It's much easier to source a well-made GNS that you can turn into gin. The truth of the matter is the real art of making gin lies in your use of those botanicals."

### Spirit Selection

Yet while Clay Smith, distillery manager at Corsair Distillery in Bowling Green, Kentucky, agrees that the selection of botanicals is key, he also seeks out GNS's based on his final flavor aims. For example, a potato vodka will favor one botanical mash bill, he says, while a grape vodka will favor another. Therefore, he advises home distillers to first go to liquor stores to research ingredient labels on vodka or pure grain spirit bottles to determine what flavors might be apparent in those distillates before compiling their own mash bill.

"There isn't one GNS that works for all gins," Smith says, adding that despite the claim that vodka is supposed to be generally flavorless, if you compare a variety of them side-by-side, they will always taste different. "You start with the basic vodka to learn to make gin if you want, but as you get better at it, you'll see that choosing the right GNS is a good way to improve your product."

Once you've created or sourced your GNS, you'll need to water it down to between 80 proof and 90 proof. This is because oils in some of your botanical mix are more water soluble, while others are more alcohol soluble. That allows both liquids to pick up and carry over every accessible flavor to the final distillate.

### Steeping and Infusion

While in whiskey making the preparation of flavor grains is pretty straightforward, there are a variety of ways to prep botanicals for gin making. According to Tripp Stimson, master distiller at Kentucky Artisan Distillery in Crestwood, Kentucky, they can be steeped in the liquid before distilling or suspended in a vapor basket in a column still for passive flavor capture. Some distillers leave botanicals whole or grinding them so their volatile oils can be accessed more easily.

"My advice is to steep the botanicals because it's the easiest way to do it for a small distiller," Stimson says, adding that steeping (also called macerating) sidesteps the need for a still with a column and vapor basket. "When you steep, you have a clear idea of your weight (of botanicals) to volume (of alcohol), plus you know how long you've let it steep before you've distilled it. That generates a baseline you can repeat or change very easily."

When Stimson consults with clients wanting a gin for large-scale production, he discusses their botanical preferences and produces separate and small distillation runs centered on each botanical. Finished liquids are then tasted and blended to the client's desires, leaving Stimson to backtrack and create a gin recipe based on the preferred amount of botanicals.

"Once we get to a blend the customer likes, I'll go back to the actual weights of the included botanicals and steep those to make one sample that gets compared to the blended sample," Stimson says. Spoken like the trained chemist he is, Stimson adds, "It's really labor intense, but you're running a controlled experiment that has to be precise."

Stimson says all distillers are prone to "overthinking gin", as it's so easy to do. "You think the recipe comes down to $2 + 2 = 4$, but it never does," he says. "So while the sky is the limit, the biggest point is to not get too crazy with it. It'll take the fun out of it."

There is a downside to steeping botanicals, as Colin Blake (DSE) points out. "From an ease of use perspective, I like a vapor basket because it's way easier to clean," he says. "To be fair, we've never done a side-by-side maceration versus vapor distillation to see how they differ. But since there are tons of vapor-basket gins out there and they taste great — and it's way easier to clean those out than to clean out your cooker — I'm going with the vapor basket."

**ABOVE:** *Juniper berries in their natural form, originally used as diuretic medication before it was used to make gin.*

When asked why gin making appears so difficult when, 300 years ago, it was widely and easily made at home in England and Europe, Blake says the not so obvious difference is the quality of the gin made by amateurs then compared to gins made today.

"If you're talking about people doing it at home, what they were probably doing was making a more neutral vodka and adding botanicals that only made the gin taste better than the water that was around," Blake says. "We're talking about the days of bathtub gin, when people literally filled a bathtub with neutral grain spirits and soaked botanicals in that. It wasn't highly scientific, and, from what we can tell, not particularly good either. Thankfully those days are long gone."

# NEGRONI

*Famous, flavorful and easy to make — Negroni is named after Count Camillo Negroni, who, while in a bar in Florence, Italy, ordered an Americano (Campari, sweet vermouth and club soda) but asked that gin be substituted for soda. Needless to say, the cocktail's potency increased, but so did its ability to counter the rich Campari.*

## INGREDIENTS
1½ oz (45 ml) gin
1½ oz (45 ml) sweet vermouth
1½ oz (45 ml) Campari
Orange peel to garnish

## METHOD
Fill a rocks glass with ice, add all the liquids and stir for about 15 seconds using a bar spoon.

To garnish, peel a thin strip of orange rind using a vegetable peeler. Fold to express its oils into the glass.

# RAMOS GIN FIZZ

*New Orleans gets a lot of credit for cocktails, and this is another sipper from the Crescent City made in the 1880s by Henry C. Ramos. Several sources claim the original recipe called for it to be shaken for 12 minutes to froth the egg white. Fortunately, this recipe gets the job done in about a minute.*

## INGREDIENTS

2 oz (60 ml) Old Tom gin
1 oz (30 ml) heavy cream
1 egg white
½ oz (15 ml) lemon juice
½ oz (15 ml) lime juice
2 tsp sugar
3 dashes orange flower water or essence
  (orange bitters will work in a pinch)
1 oz (30 ml) club soda

## METHOD

Half-fill a small shaker with cracked ice and add all the ingredients except the club soda. Cover and shake vigorously for 1 minute. Using a Hawthorn strainer, strain the mixture into a chilled rocks glass and top it up with soda for a gentle fizz.

# WORDS OF WISDOM

## CLAY SMITH
### DISTILLERY MANAGER, CORSAIR DISTILLERY

Bowling Green, Kentucky and Nashville, Tennessee
www.corsairdistillery.com

Clay Smith is the distillery manager at Corsair Distillery in Bowling Green, Kentucky and nearby Nashville, Tennessee. Under the direction of Corsair co-owners, Darek Bell and Andrew Webber, Smith oversees the production of what his bosses dub "hand crafted small batch ultra-premium booze for badasses." Their award-winning gin is remarkably bright and fragrant, a spirit that turns an ordinary gin and tonic into the topic of discussion at a party. Corsair's wide array of limited release whiskeys, rums and vodkas are fantastic, too, but we'll stick with gin here.

**Some assume that all clear spirits are simple to make. I understand that's not the case with gin.**

In so many cases, distillers like us are taking neutral spirits and adding botanicals to it, so we're not dealing with fermentation, which makes it easier. But gin definitely has its challenges. You can very easily slip off the edge and create a botanical mess and ruin the spirit. There's an art to getting the right botanical balance.

**Provide us with an example of that.**

Lavender is the scapegoat example we go to when we want to demonstrate this in a class. So many have this idea that they want a gin with a floral flavor profile, something that has a very good bouquet, and so they think lavender. But when they use it, they realize it doesn't translate in the fashion they think it will; its essences come off in a weird way, smelling like feet or really bad cheese — or really good cheese, depending on your cheese palate. The point is it doesn't have the desired effect, and what a lot of people at that level don't understand are the chemical processes that change that essence to something less than desirable.

**There is such a broad range of gin-favorable botanicals. So how do you determine what you might want and adjust the proportions for your recipe?**

Starting with a recipe you like and then adjusting yours to your tastes is one way. But if you're doing it precisely and scientifically, you're doing separate runs using the exact same amount of each botanical in each run. Then you're blending (those distillates) proportionally to create a compound. This way you have a lot more control. Lots of distillers are making gin that way successfully. Trying to adjust the amounts of a load of botanicals with each run is more complicated.

🍶 **What about buying gin flavoring and adding it to your distillate?**
We would not recommend that, mostly because it's not our method of doing things. And I don't think many home distillers we talk to are doing that either. In my experience, if you're a true believer in the process, then being involved in those details is something you don't want to skip. Some of the folks who come in and say they're playing around with a home still are already using their own blends based on their own needs. I honestly don't know of anyone using flavoring (in home distilling).

🍶 **Since making high-proof gin from your own ferment is so costly and time consuming, do you recommend any particular 190 proof grain-neutral spirit (GNS) off the shelf?**
It's not really a one-type-fits-all situation, for sure. I would pay close attention to what the base spirit was originally made from because some botanicals work better with different GNS's. At Corsair, we want the best solution for each specific botanical load. We're told that since vodka is distilled to 190 proof that it's neutral in flavor. But if you tried 10 of them side by side, you'd taste differences in all of them.

So when you make your gin, experiment: see what a potato vodka does with a particular botanical load or another, and the next time, see what a barley or grape based-spirit does. It changes so much and that process of deciding what you like is the fun part.

🍶 **Talk about the importance of reducing the GNS's proof before distilling a batch of gin.**
We actually cut our 190 proof down to 95 or 100 before we start. Some of the flavors in our botanical mix that we want to come off in the distillate are adhering to water molecules as well, so you want that

**ABOVE:** *Corsair gin at the distillery in Kentucky.*

in there too. Using it straight off the shelf isn't the best way to get the most flavor.

🍶 **In classes you teach on gin making, do you tell your students to steep their botanicals, or to place them in a gin basket?**
We do use a basket at Corsair, but I teach maceration also, so students can see the difference between the two techniques and make the decision themselves. We find in our own practice that the vapor basket trumps all, but that may not be true for the type of spirit you want to produce or the type of still you have. We do know that it's a lot easier to unload a vapor basket once you're done than it is to strain and clean out the still.

For those who like maceration, we have seen that more grain notes came through when we did a genever product started from a low wine. Was it a lot? Not really. But it's safe to say there was a little difference depending on whether we used a vapor basket or the maceration process.

# OLD MAID

*If you're a fan of gin and already like a mojito, you'll love this light cocktail with the added freshness of lime juice and cucumber. The key to a good Old Maid is to muddle gently, as crushed mint can become bitter. I'd also recommend using Hendrick's gin.*

INGREDIENTS

2 oz (60 ml) gin
1 oz (30 ml) lime juice
¾ oz (22 ml) simple syrup (1:1 sugar to water)
3 thin cucumber slices
4–6 mint leaves
1 small mint sprig, to garnish

METHOD

Add the cucumber slices and mint leaves to a small shaker and muddle gently, just until you smell the mint aroma. Half-fill the shaker with cracked ice and add the gin, lime juice and simple syrup; cover and shake vigorously for about 5 seconds. Fill a rocks glass with ice and strain the mixture into the glass. Garnish with the mint sprig.

# TOM COLLINS

*Its year-round versatility has made this one of the most popular gin cocktails in the world. For the sweetener, some mixologists add complexity with infused fruit syrups. Since club soda is used, I find London dry gin gives the cocktail bite, but Old Tom is a nice variant.*

INGREDIENTS

2 oz (60 ml) gin
1 oz (30 ml) freshly-squeezed lemon juice
1 tsp (5 ml) simple syrup (1:1 sugar and water)
3 oz (90 ml) club soda
1 maraschino cherry
1 orange slice, cut into two halves

METHOD

Fill a Collins glass with ice and add the gin, lemon juice and simple syrup; stir for about 15 seconds to chill. Add the cherry and one of the orange slices and top with club soda; stir gently to blend. Garnish with the second orange slice.

**RIGHT:** *Tom Collins*

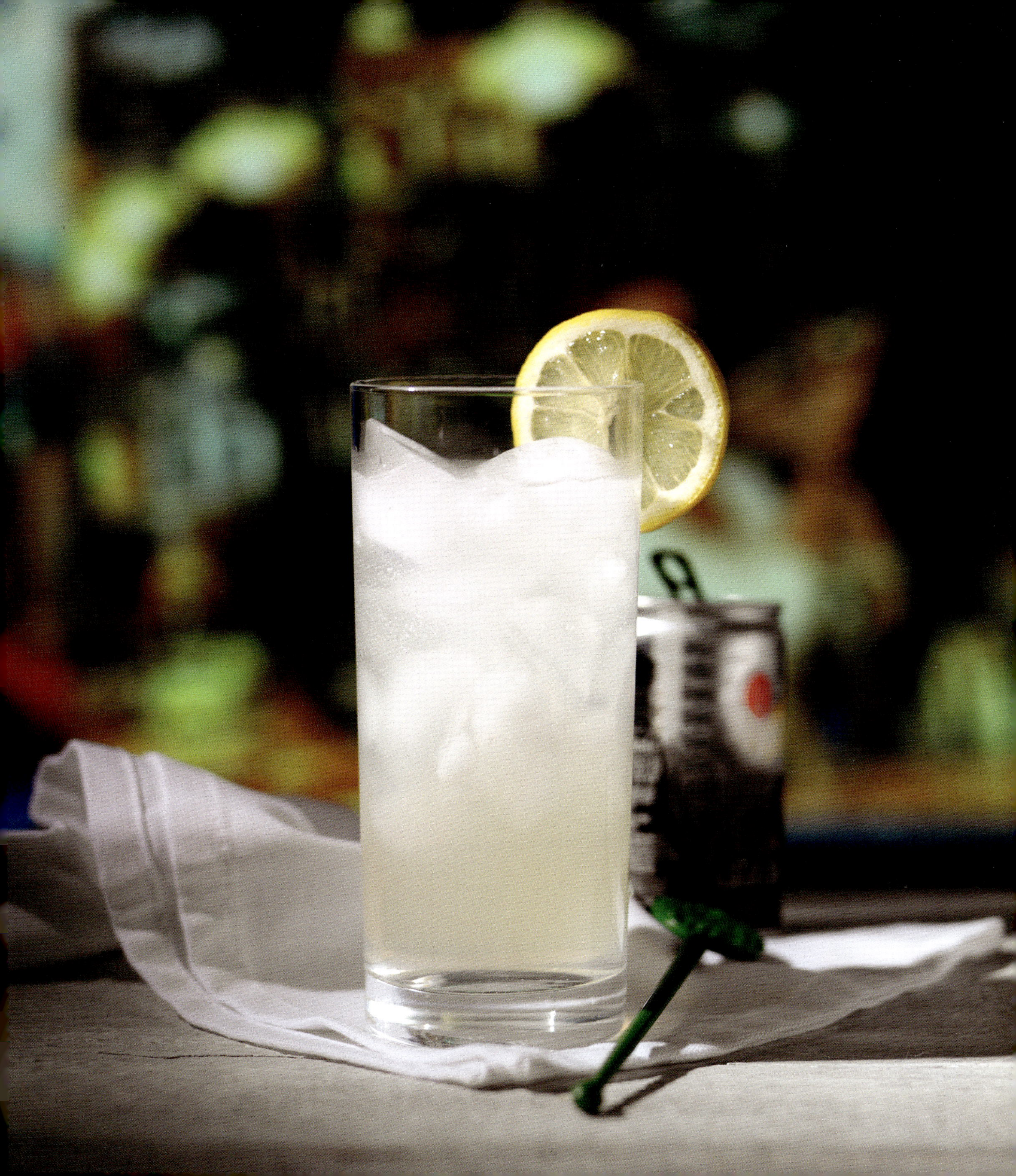

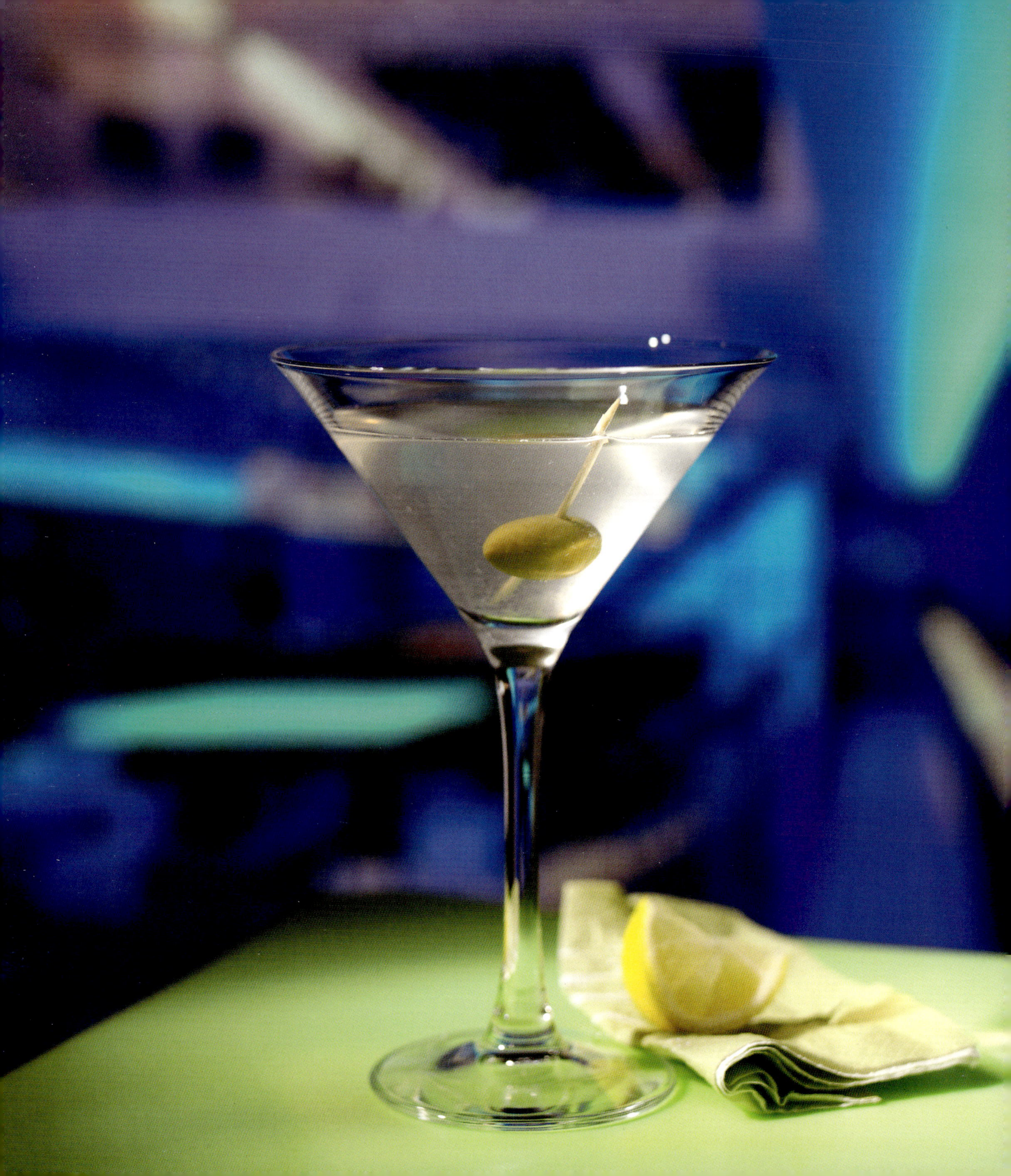

# MARTINI

*Few cocktails are simpler to make, yet how to manage the few ingredients in a Martini has evoked much debate. Shake or stir? Use all gin and just a dash of vermouth? Shake only the gin and rinse the glass with vermouth? The only thing all agree on is it must be served cold.*

INGREDIENTS
2 oz (60 ml) gin, preferably London dry
1 oz (30 ml) dry vermouth
1–3 olives skewered or lemon peel twist, to garnish

METHOD
Place a coupe or up glass in the freezer 15 minutes before you start. Shaken martinis contain tiny bits of ice that slip through the strainer. If you prefer all liquid, choose the stirred version. For a "dirty" martini, add cold brine from the olive jar to the drink.

If you stir: Fill a mixing glass or small shaker with ice cubes, add the gin and dry vermouth and stir for 30 seconds using a bar spoon. Strain into a chilled coupe or up glass using a Hawthorne strainer, and garnish.

If you shake: Half-fill a small shaker with cracked ice and add the gin and dry vermouth; cover and shake vigorously for 5 seconds. Strain and garnish, as above.

**LEFT:** *Martini*

# VIEUX MOT

*This cocktail calls for Plymouth gin, which is less dry, and lighter overall on botanicals. Its less-assertive flavor means, as bartenders say, "it plays well with other ingredients in the drink." This one uses St. Germaine, a vibrant elderflower liqueur.*

INGREDIENTS
1½ oz (45 ml) Plymouth gin
¾ oz (22 ml) St. Germaine
¼ oz (7 ml) simple syrup (1:1 sugar to water)
¾ oz (22 ml) freshly-squeezed lemon juice
Lemon peel, to garnish

METHOD
Half-fill a small shaker with ice and add all the liquid ingredients; cover and shake vigorously for about 5 seconds. Strain into a chilled glass using a Hawthorne strainer.

To garnish, peel a thin strip of lemon rind using a vegetable peeler and twist to express its oils into the glass.

# LAWN DART

*Claiming there's just one way to make a Lawn Dart is about as dangerous as using those infamous toys themselves. Either way, someone could lose an eye! I found, that the liquors in recipes were clear, not brown, and I've settled on equal parts of gin and tequila.*

INGREDIENTS
1 oz (30 ml) London dry gin
1 oz (30 ml) blanco tequila
¾ oz (22 ml) freshly-squeezed lime juice
¾ oz (22 ml) agave syrup
¼ oz (7 ml) green chartreuse
2 small squares of green bell pepper
1 lime wedge, to garnish

METHOD
Add the agave syrup and bell pepper pieces to a small shaker; muddle both, gently mashing the pepper. Fill the shaker halfway with cracked ice and add all the other liquid ingredients; cover and shake vigorously for about 5 seconds. Fill a Collins glass with ice and strain the mixture into the glass using a fine mesh strainer. Garnish with the lime wedge.

# CAPRICE

*This old-school cocktail is making a comeback not only for its simplicity, but its openness to barrel aging. At the 1886 bar at The Raymond Hotel in Pasadena, California, they're aging the cocktail in used whiskey barrels from four months to a year and selling flights of Caprices.*

INGREDIENTS
1½ oz (45 ml) London dry gin
1½ oz (45 ml) dry vermouth
½ oz (15 ml) Benedictine
1 dash orange bitters (Regan's is recommended)
Orange rind, to garnish

METHOD
Half-fill a small shaker with cracked ice and add all the liquid ingredients; cover and shake vigorously for 5 seconds. Strain into a chilled coupe or rocks glass.

To garnish, peel a thin strip of orange rind using a vegetable peeler and twist to express its oils into the glass.

**RIGHT:** *Caprice*

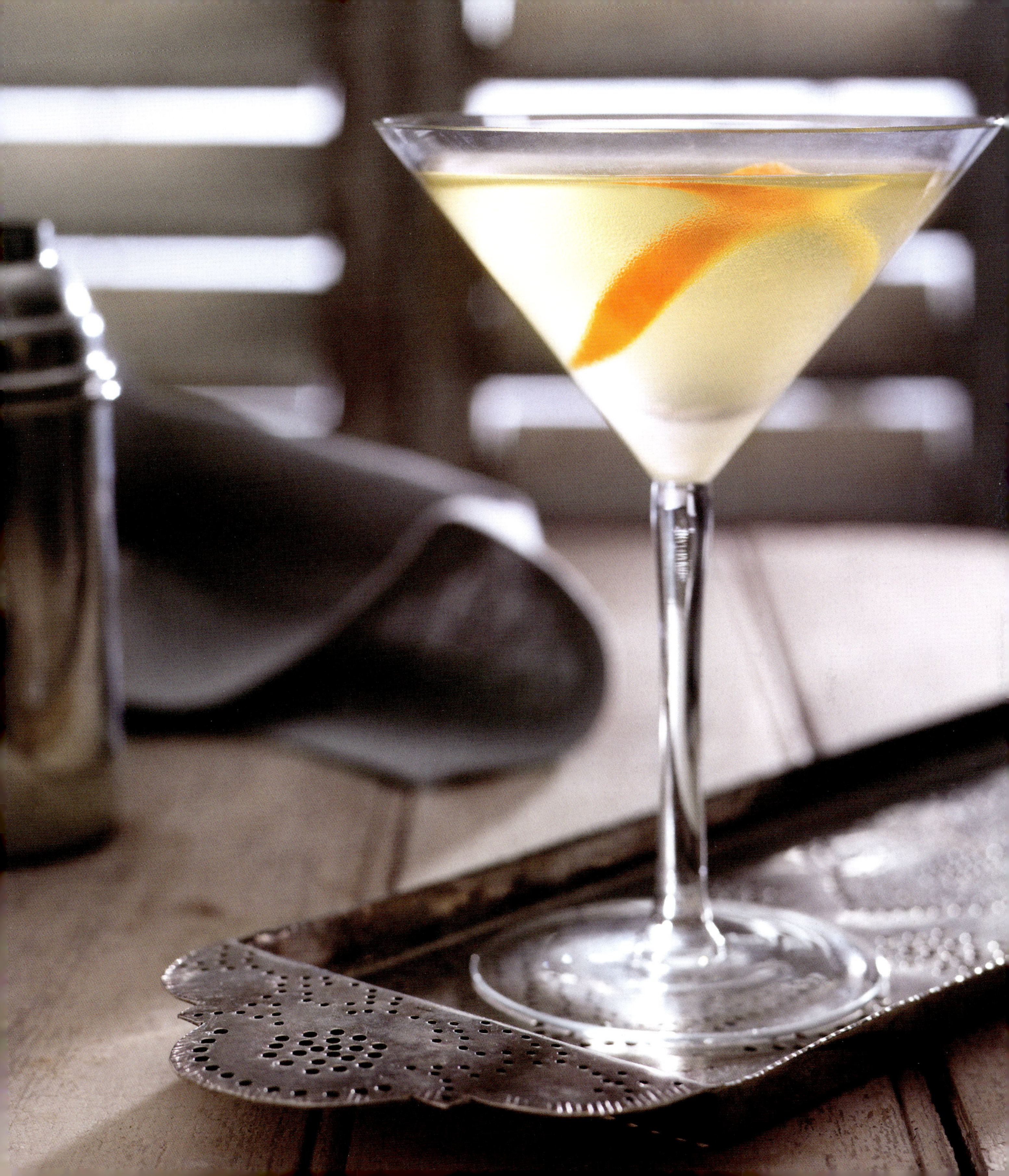

# WORDS OF WISDOM

In 2012 he partnered with Lewis Hayes to found London Bar Consultants. The two men have trained countless staffers in the reopening and relaunching of numerous drinking venues and experiences, while helping to bring large and boutique brands to the market. Suffice it to say, Brown is a gin fanatic and expert. On the shelves at LBC's Merchant House bar are more than 250 premium, international gins. In early 2015, Merchant House also was the site for Think Gin!, a trade event that attracted some of the biggest industry names for education and inspiration.

### How and when did you become interested in spirits?

I remember my father drinking Bacardi on odd occasions, so being a curious kid I sneaked into the drinks cabinet and had a sip from the bottle. It was so awful that at first I couldn't imagine why he was drinking it, so I set about researching and trying to understand it.

### Why are you so passionate about spirits?

Because there's more to a drink than what goes in the glass. The taboo is such a shame given how rich and colorful the history of spirits can be, and how much value there is to be found in the nuances between liquids.

### Which (if any) is your favorite spirit and cocktail?

The one in my glass.

### Why has gin seen a recent revival?

Why a revival or why recently? For a product that is relatively easy to make given the infrastructure of spirit and botanicals, and its potency in scope of flavor,

## NATE BROWN
### PARTNER, LONDON BAR CONSULTANTS

London, UK
www.londonbarconsultants.co.uk

Nate Brown began his cocktail-making career in Manchester, England, where he became a company trainer for a successful group. He became a key figure in the bar renaissance of the city's Northern Quarter before moving to London to ply his trade in some of the capital's most respected venues.

**ABOVE:** *The Merchant House Bar in the City of London, where Nate and his team serve a vast array of gin and rum.*

it was only a matter of time. The drinks industry is no different to any other in this regard: the consumer is interested in value. For gin this perhaps comes from the unique botanical blend, which is achievable by any budding distiller, or more likely in the back story explaining "why" this gin exists. It has never been easier or more necessary to communicate one's unique selling proposition.

### What circumstances and environment are best for enjoying the perfect gin?

The flexibility of the spirit is unmatched. It can be a savior or a solace, a pick-me-up or nerve settler, elegant and refined or deep down and dirty.

### What does the future hold for the distilling industry?

I expect the market to become less forgiving. Everyone talks about the liquid and the potential for the liquid. But the truth of the matter is that it is the selling that really counts. Expect more correlation between the branding and the liquid, and more innovative ways of reaching the consumer. Who will succeed? Only time will tell.

CHEMY
RY GIN
Fields of Maine
Pure Grain
RTLAND, MAINE
OL. (90 PROOF)

# CRAZY TRACY

*If you don't want to wind up as crazy as the guy after whom this drink was named, respect its potency. It was created in an effort to use a bottle of spiced gin, but you can mimic the spice profile by adding equal parts of Peychaud's and Angostura bitters to regular gin.*

INGREDIENTS
2 oz (60 ml) gin
½ oz (15 ml) St. Germaine
½ oz (15 ml) Cointreau, or any orange liqueur
4 oz (120 ml) freshly-squeezed orange juice
½ oz (15 ml) pomegranate juice
3 dashes Angostura bitters
3 dashes Peychaud's bitters
1 thin half orange slice, to garnish

METHOD
Half-fill a small shaker with cracked ice, add all the liquid ingredients and blend by stirring with a bar spoon for about 15 seconds. Using a Hawthorne strainer, strain the mixture into a Collins glass filled with ice. Garnish with the half orange slice.

**LEFT:** *Crazy Tracy*

# FRENCH 75

*The name of this drink is claimed to come from its potency; it's said to have the kick of a 75mm French artillery gun, although it's actually quite mild. While the mimosa and Bloody Mary may reign as American brunch drinks of choice, I'd recommend this with equal enthusiasm.*

INGREDIENTS
1 oz (30 ml) London dry gin
½ oz (15 ml) lemon juice
½ oz (15 ml) simple syrup (1:1 sugar and water)
1 oz (30 ml) champagne
Lemon rind, to garnish

METHOD
Half-fill a small shaker with cracked ice, add the gin, lemon juice and syrup and shake vigorously for about 5 seconds. Strain into a chilled coupe glass or champagne flute using a Hawthorne strainer and top with champagne or good sparkling wine.

To garnish, peel a thin strip of lemon rind using a vegetable peeler and twist to express its oils into the glass.

# BRANDY

*Brandy is defined simply as an alcoholic beverage distilled from wine or a fermented fruit mash. And that's where the simplicity ends. Depending on the country in which it's made, the region of that country in which the fruit was grown, the fruit that was mashed to make it, or the wine that was rectified in the still; brandy not only takes on a vast number of names, but a delicious diversity of representations.*

RIGHT: *Twice pot-distilled Cognac, widely regarded as one of the finest of brandies.*

```
Claret is the liquor for boys; port for men; but he
who aspires to be a hero must drink brandy.
                              - Samuel Johnson
```

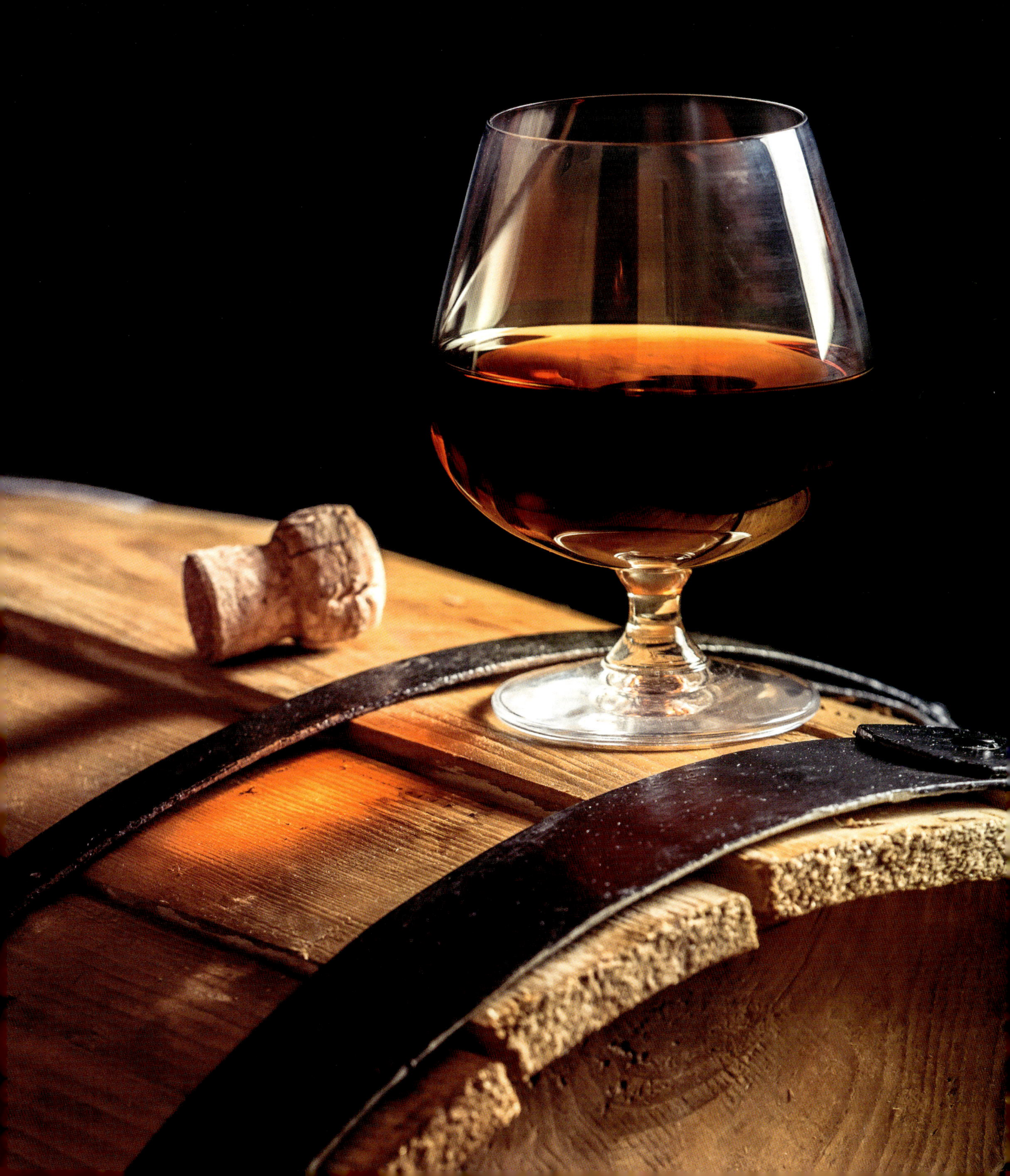

## A LITTLE HISTORY

As distillation spread throughout Europe in the 15th century, its adaptation to wine was at first a bit misunderstood. The thought process was this: heating would reduce wine's water content, concentrate it and make it lighter for transportation; once the reduced wine arrived at its destination, it could be reconstituted and consumed. What those early alchemists didn't anticipate was how dramatically heat changed the wine's chemical composition. (The Dutch called the resulting product bradwijn, meaning "burned wine.") Their experiments weren't for naught, as what emerged from the still was not only quite appealing, it improved markedly after a rest in wood casks, the main storage vessel of the day.

While some may argue which country's brandy is best, French brandies are widely regarded as the finest of the craft, and finest of all are Cognac and Armagnac. Cognac, the twice pot-distilled spirit named after the town in which it was born, was first distilled in the 17th century. It is made only from grapes grown in regions surrounding the town and wine produced there. It's aged in new oak casks for a year before being transferred to used cooperage for another rest of 2 to 15 years. Armagnac, which is made in the French county of the same name, is created from a four-grape wine that's distilled once in a column still. It's commonly aged for 10 or more years, though extended aging runs 20 to 30 years.

Brandy migrated west to the U.S. in the 1800s as Spanish missionaries began making brandy in California, where the country's largest producers still operate.

The century before this, some 150 years before in fact, pisco was created in the Peruvian port of Pisco. It ranges in color (clear to golden) and varies from 60 proof to 100 proof. It is the base spirit in the famous pisco sour cocktail. Chileans also distill pisco, but their right to do so is the source of an ongoing dispute with neighboring Peru.

In general, brandies of many types are produced all over the world, especially where grapes and other stone fruits grow easily, such as in Germany, Greece, Spain, Italy, Portugal, Australia and South Africa.

## TYPES OF BRANDY

Armagnac and Cognac are undoubtedly the most influential of brandies, but let's visit some other popular brandies and take a look at their standards of identity:

**American brandy.** Must be aged in oak for at least two years or labeled "immature." If it's not distilled from grapes, the label must specify the fruit source, for example "peach brandy."

**Applejack.** Made from American hard cider that's commonly blended with neutral spirits; depending on the distiller, it may be aged in wood.

**Calvados.** Made in France from apples and pears and aged at least two years.

**Eau de vie.** A French term for unaged grape brandy that also applies to unaged brandies made from fruits other than grapes.

**Grappa.** Made in Italy from grape pomace; most grappas aren't aged, but the practice of barrel resting the potent potable is increasing.

**Kirsch.** Made in Germany from cherries but not aged.

**Marc.** The name given to grappa when made in France.

**Schnaps.** The German term for unaged grape brandy and unaged brandies made from fruits other than grapes — not to be confused with "schnapps", those cheap bar shot favorites.

### Brandy Ratings

If you know and understand how brandy is rated, you're probably thankful brandy distillers go to the trouble of clarifying what you're getting. If you're new to the rating system, you might find it a bit overwhelming, especially if you're trying to master it while under the influence of brandy! Let's begin with the letter system itself:

C = Cognac
E = Extra
F = Fine
O = Old
P = Pale
S = Special
V = Very Special

**ABOVE:** *Grape vines in the Cognac region of France.*

**ABOVE:** *Cognac distillery in Chareute, France.*

Sometimes the letters are co mbined to form a new designation, while at other times they stand alone. Here's what they mean:

- VS = Very Special, aged at least two years (also called "three star")
- VSOP = Very Special Old Pale, aged at least four years (also called "five star")
- XO = Extra Old, aged at least six years
- E = Extra, aged six years or more

Here are a few other notable terms outside the brandy "alphabet":

**Hors d'Age.** Beyond age, meaning it is too old to determine its age.
**Napoleon.** Brandy aged at least four years.
**Vintage.** Age of brandy determined by the actual time spent in the cask prior to bottling.

## DISTILLATION

As the distiller, you have many options on where to begin: you can crush your own fruit, or ferment prepared juice, or distill from wine. While pressing pears or apples or crushing grapes may sound romantic and true to the craft, it requires real labor, an investment in presses and a lot of tool cleaning.

For his apple jack brandy, Alan Bishop, assistant distiller at Copper & Kings American Brandy in Louisville, Ky., has crushed whole fruit and fermented prepared juice. Not surprisingly, he prefers buying juice, which he then fortifies with brown and white sugars. Others suggest finding a nearby apple orchard where you can buy several gallons of fresh cider.

Your yeast choice is also significant since it will influence the flavor you'd like to achieve, and it should be matched to the fruit you'll be fermenting. Among commonly recommended yeasts are English cider and sweet mead yeast. When he was a hobby distiller,

Bishop preferred champagne yeast because it's faster acting than baker's yeast: "It works well, but the flavor leans a bit more toward fruity esters," he said. "Bread yeast is great for flavor, though it can take longer to ferment: about one to two weeks at 70°F (21°C)." (Note: Since brandy washes and mashes are fermented at significantly lower temperatures than grain whiskies, fermentation lasts much longer. So plan that extra time into your process.)

At the professional level, yeast selection blends science and art, says Bishop's boss, head distiller Brandon O'Daniel: "Yeast is super-important with brandy because it's so specific to the grape variety. If I'm fermenting a Chardonnay, it'll have one type of yeast, and if I'm fermenting Cabernet Franc, it'll have another type of yeast. I work with probably 15 to 20 different types of yeasts based on what I want to pull out of that wine."

If you're using a pH meter to check your wash or mash before pitching the yeast, grape juices will typically range between 3 and 4, while apple juices run closer to 3. (By comparison, the average grain mash has a pH of 5.5.)

Aerate your wash by transferring your mash back and forth a few times from your fermenter to a similarly sized bucket. If your mash temperature is 70°F (21°C), it's safe to pitch the yeast. O'Daniel gets very hands-on at this stage. "If I'm working with less than a ton of grapes, I'll just stick my hand in (the crushed grapes) and swish it all around. But if it's more, I'll use a paddle," he says. "I do like to touch it. There's something about sticking your hands in that makes you feel part of what you're creating."

The abundant solids from the crushed grapes will float to the top and form a cap, and O'Daniel adds, "You've constantly got to push that cap back down three to four times a day during the first two weeks (of fermentation). Don't agitate it a lot, but do break up that cap."

Ferment the mash or wash until "dry," meaning all fermentable sugars are converted to alcohol. You can measure this by using a hydrometer, which should read 1.0.

**ABOVE:** *Brandy barrels in storage, Midi-Pyr, France.*

While bacterial growth is always a concern when fermenting wine, O'Daniel says he never uses sulfites to eliminate that risk: "Every flavor you put in becomes magnified by a factor of 10 during distillation, so if you add something that tastes like sulfur, you will get 10 times the sulfur flavor in the spirit." He continues, "People don't always seem to believe that you can't turn crappy wine into great brandy. It's been tried before, but it does not happen."

Hobbyists and experts interviewed say they prefer pot stills over column stills for brandy distillation, claiming the spirit is fuller bodied and that fruit flavors are carried over better to the end product. Where they disagree slightly is on straining off the lees (crushed, fermented fruit) before distillation. Some say adding the

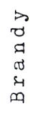

**ABOVE:** *"Rakia", or traditional brandy, being distilled undercover in the village of Cherven Brayg, Bulgaria.*

lees to the kettle during the stripping run yields more complex fruit flavor, while others, such as O'Daniel — who has a 500-gallon (nearly 3,000 l) still to clean — prefer the lees removed.

On average, O'Daniel says stripping runs produce a 70 proof spirit, but that the subsequent finishing run will turn out 135 proof distillate. At Copper & Kings, that's barrel proof. The distillery waters the aged brandy back to 90 proof for bottling, which is a good bit higher than the common industry standard of 80 proof.

## AGING

Brandy can be delicious unaged, but it clearly benefits from a barrel rest. Copper & Kings American Brandy (see page 126) likes used bourbon barrels with a medium toast and a No. 2 char. Its brandy rests four to six years in its climate-controlled basement.

But since few hobbyists have the budget or space for cooperage, adding toasted oak chips to the stored distillate will add wood flavors. Just remember to taste the spirit about once a month to see if it's achieved the flavor and color you want. Once there, filter it (carbon filters are widely available for the task), bottle it and enjoy.

"The important thing to remember about home distillation is that it is not necessarily about achieving perfect consistency," says Bishop. "It's about constant refinement and experimentation." In other words, enjoy the whole process, and the hard work involved, as much as the results.

# BRANDY AND GINGER

*If you think this is just another ginger highball, think again. This simple setup chooses immature brandy, as in ne'er touched by a barrel brandy — the stuff you'll be making. The leaner, fruitier spirit delivers an array of great flavors. For more bite, use ginger beer.*

## INGREDIENTS
2 oz (60 ml) immature brandy
Ginger ale
1 lime wedge, to garnish

## METHOD
Fill a chilled cocktail glass with ice, add brandy and stir for 5 seconds using a bar spoon. Top up with ginger ale to taste, and garnish with a squeeze of lime.

# BRANDY ALEXANDER

*There's a version of this drink made with vanilla ice cream. While it's tasty, the addition of ice cream turns a well-structured drink into dessert — not ideal if you've spent time making your own brandy. Though still richly textured, this leaner version is a delight.*

## INGREDIENTS
2 oz (60 ml) aged brandy
¾ oz (22 ml) crème de cacao
   (Tempus fugit is recommended)
¾ oz (22 ml) heavy cream
Freshly-grated nutmeg, to garnish

## METHOD
Half-fill a small shaker with ice, add all liquid ingredients and shake vigorously for about 5 seconds. Using a Boston shaker and fine sieve, strain the drink into a chilled coupe and garnish with the freshly-grated nutmeg.

**RIGHT:** *Brandy Alexander*

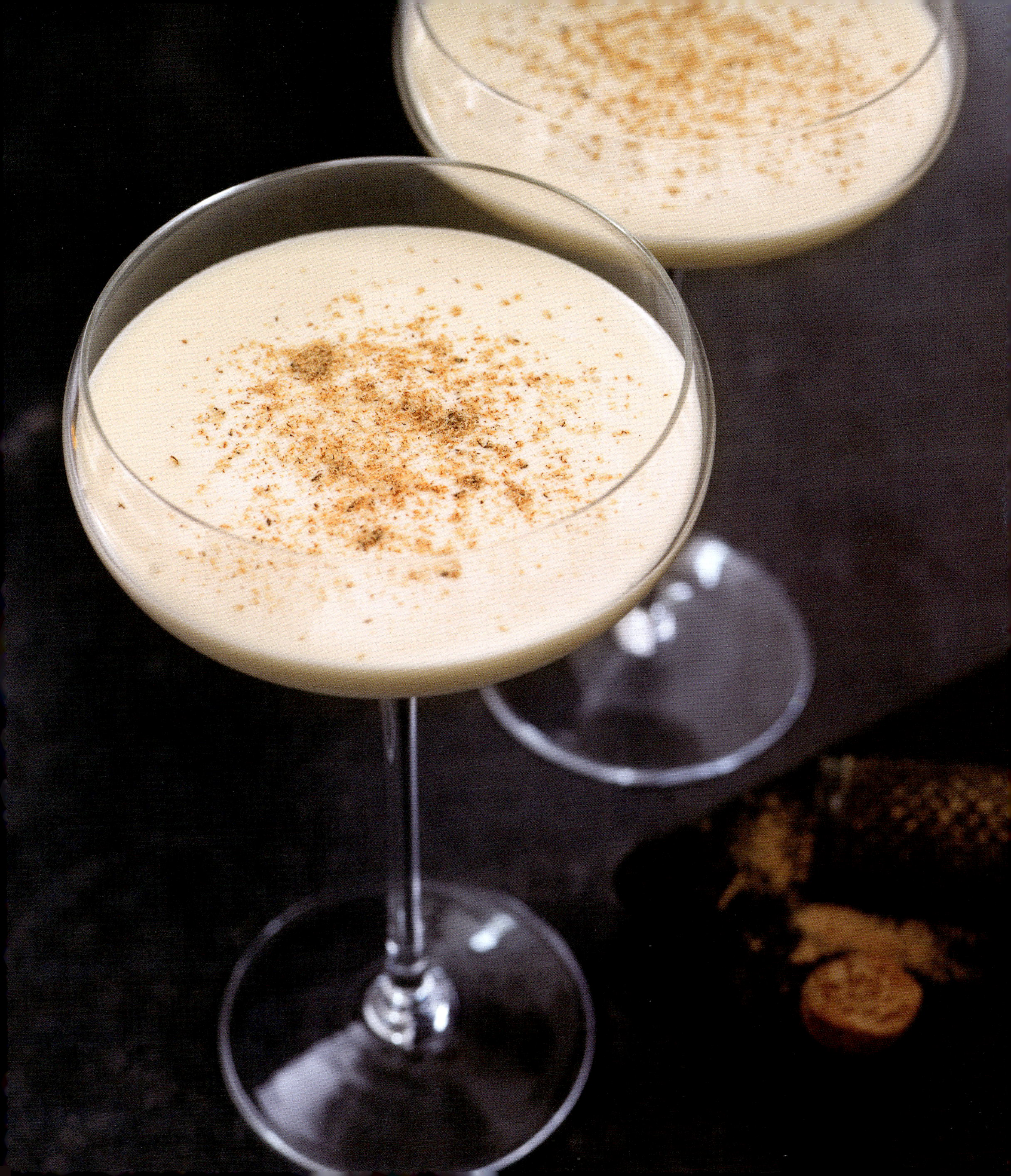

# WORDS OF WISDOM

## BRANDON O'DANIEL
### HEAD DISTILLER, COPPER & KINGS AMERICAN BRANDY
Louisville, Kentucky
www.copperandkings.com

When the owners of Copper & Kings sought a Kentucky source for wine they could distill into brandy, they chose Elk Creek Vineyard. And when they discovered Elk Creek winemaker Brandon O'Daniel also knew a thing or two about distilling, they lured him to their operation in 2014.

**How did you make the jump from winemaker to distiller?**

(Owners) Joe and Leslie Heron were looking for someone to produce bulk wines for them, and they'd had my wines before and liked them. Over lunch one day we kind of hit it off, and they called back later that day and asked if I'd be interested in interviewing for the job of distiller. I was, and when I got there, I saw the stills and was pretty much hooked. I've not left since.

**Seems every hobbyist makes vodka, and some do grain whiskey, but not a lot are doing brandy. Why do you think that is?**

It's all fairly similar at the still, but fermentation takes so much longer for brandy. With a simple grain mash, you can get it fermented in three or four days and have it ready for the still. With grapes, it takes close to a month. Unless you're just going to buy wine or juice and distill that, you've got to come up with grapes, crush and press them and then turn that into wine, which we do. I don't think too many people are interested in going to all that trouble at home, or maybe they do and we just don't know about it.

Of course it was different for me since my family was in the wine business. We did a little distilling because it was fun and because it was just another alcohol to make since we always had the materials to do grappa or brandy. We'd make a little for ourselves in a 15 gallon (56 l) still, but that was a long time ago. I've got these now (he gestures toward C&K's brandy stills).

**What are the major differences between working on small versus large stills?**

Basically it's scale, plus a few more bells and whistles. There's also lots of piping to move distillate around the building, and there's a lot more money at risk. It's one thing to run 5 gallons (20 l) of spirits for grappa from must (by-product of grape crushing) you were going to toss out anyway, but it's another to waste a 1,000 gallon (3,785 l) run here. Thankfully, that's not happened.

When your runs are that large, it's a matter of slowing the process down and doing it right. We spend a lot of time on the stills. We take temperatures on our final runs every 10 minutes and on our stripping runs every 15 to 20 minutes. We're constantly tasting and checking our hydrometers, and we're taking a tremendous amount of notes.

Taking lots of notes helps, especially when you're making multiple types of alcohol. Having those notes prevents a lot of backtracking because you can go back and look at the last three months of notes of every distillation run and get an idea of what went right. It allows that next run to go a little smoother.

### So say you find the hobbyist who wants to do brandy, how would you coach them through it?

I'm like everybody else: I'd say, "If you're doing it, we're not condoning it." But if I did talk to them, I'd ask them what they like to drink and start from there. I'd tell them to keep their still setup simple, to start small, but to use at least a 5 gallon (20 l) still. You can do brandy on those small 1½ gallon (6 l) stills, but doing such a small amount, it makes it super hard to separate your cuts cleanly and get a feel for the distillation cycle. Making a larger amount gives you more latitude in your cuts.

But instead of making brandy, I'd have them start by making vodka. Or they could do a really simple brandy (from finished wine) before getting into different grape varieties. That gets harder quickly. So just learn the actual process of running your still, get good at it, and then get into more different and complicated types of alcohol.

And I'd definitely tell them that they can't turn crappy wine into great brandy just by distilling it. It's been tried before and it does not happen. Don't let anybody tell you it will.

**ABOVE:** *Copper & Kings American Brandy 1000 gallon stills made by Vendome Copper and Brassworks.*

### Now that you're officially a legal distiller, do hobbyists seek you out for advice?

Constantly. Somebody is always bringing a jar by and asking, "What's wrong with this?" Or, "If you were fermenting wine, what would you do?" I get lots of those questions — but not so directly. You will see some interesting stuff in the parking lots at some of the trade shows we go to. I don't mind answering questions, though. I love talking about grape growing and brandy making. I could do it all day long.

# WORDS OF WISDOM

## ALAN BISHOP
### ASSISTANT DISTILLER, COPPER & KINGS AMERICAN BRANDY
Louisville, Kentucky
www.copperandkings.com

After Alan Bishop's father, Dale, taught him to distill at home, he was hooked. He loved perfecting his craft and recipes, and building larger stills. When Copper & Kings advertised its need for skilled still hands, Bishop grabbed the job. Dale joined his son for the interview.

**You were quite passionate about hobby distilling, right?**

Yeah, you could say that. I had a 200 gallon (750 l) still, but it's gone now. That had to go away when I got this job. A big reason I got into it was I saw it as holding onto a family tradition that I did not want to let go of.

**Did you know anything about brandy from your personal experience?**

I distilled some apples and used some apple juice base. I've done banana, pawpaw and plum brandy. Sometimes they were good, sometimes they were terrible. You learn as you go about what works.

**Dale:** Alan improved on what we were making before, made it a lot better because he took more pride in what he did.

**Alan:** I also paid more attention to what was going on in the process. I focused more on grain bills, temperatures, fermentation, and I learned everything I could about it. It's fun just to learn an art form and a trade and conquer it as best you can.

**Alan, what made you decide to distill for a living?**

I'd wanted to for a long time, so I basically started harassing people and telling them I wanted to be a distiller. As soon as I saw Copper & Kings wanted help, I started harassing them about getting an application. After about the fourth e-mail I sent, they told me to send a résumé. I just wasn't going to give up.

**Share some quick tips with anyone who's considering distilling at home.**

I'm not telling anybody to do this, but ... if you do ... a 10 gallon (40 l) still is the smallest you can run and really do good cuts. The larger the still, the easier that gets. If you can do good cuts on a 10 gallon (40 l) still, you can run liquor on anything.

Always pay constant attention to the temperature of the pot. Learn to feel the pot and know what is going on inside. Admittedly, it's not always exciting. Sometimes distilling amounts to a whole lot of staring at something for a long time. But when it starts running, there are a lot of fast movements. We tell our still hands here that distilling is about 60 percent hard science, 30 percent dark arts and 10 percent intuition. That last part is the part that matters. If you think something's wrong, don't second guess.

**Dale:** You get to the point that you just listen to a still and know it's running right.

**Alan:** Well, I can't really do it that way here. We take tons of notes. We make incredibly tight cuts on the stills.

**Dale:** Yeah, you've also got great equipment! We never had stuff like you've got now! He's not out in the rain anymore, or doing any of that, "Oh, the still's getting cold because the wind's blowing. Turn up the heat!" Or, "The wind's died down, turn it down."

**ABOVE:** *Still badge on one of Copper & Kings American Brandy stills.*

### Alan, is there a lot of study involved now that you're a professional?
Whether it's a hobby or a job, you should always read everything you can about this. Go to those home distiller forums to learn, and read any books you can on distillation. Not just whiskey, gin, or rum alone, try to take it all into account.

### So now that you're doing it legally, do you miss the sort of romantic notion of being a moonshiner?
That's kind of funny, that romantic allure surrounding all that stuff. But the truth is it's not romantic once you get into it. There's work involved.

Oh, another thing: Get away from using thump barrels. Those are training wheels. If you can get away from those, you will make product that's a lot better. Then, get really good at it and then try to find a job doing it legally.

### Dale, has the teacher become the pupil now that Alan is a pro?
Alan has taught me a lot, and now I know how it works. When I was doing it myself, I knew what was good liquor and what wasn't good liquor, but I didn't know how it all worked. I learned how to distill when I was in high school, so back then it was more about a cheap way to get a buzz. He always took it more seriously.

But I'll tell you the truth: Some of the best moonshine I've ever drunk came off what really looked like bad stills. Half the time I look at them and say, "You call that a still?" That liquor was made by guys who had been doing it their whole lives and they had a feel for it. They didn't measure anything, they just went by feel. That doesn't happen in Alan's job.

# THE DAISY

*Recipes for this shaken cocktail have been found in late 19th century books on entertaining, so calling it a classic is an understatement. While The Daisy is basically a sour cocktail, some variations use sweet touches, such as grenadine or other fruit syrups, to smooth off the edge.*

### INGREDIENTS
2 oz (60 ml) brandy
¾ oz (22 ml) orange liqueur
    (dry curacao or Cointreau)
¾ oz (22 ml) freshly-squeezed lemon juice
2 dashes club soda

### METHOD
Half-fill a shaker with ice, add the brandy, orange liqueur and lemon juice; cover and shake vigorously for about 5 seconds. Strain through a Hawthorne strainer into a chilled rocks glass and top with two dashes of club soda.

# THE AMPERSAND

*Who knows why it's named after the "and" symbol? But the "and this, and this, and this" mnemonic certainly makes its portions easy to remember. Big-bodied Old Tom gin adds heaps of personality to an old style cocktail. If you don't like gin just double up on cognac.*

### INGREDIENTS
1 oz (30 ml) cognac
1 oz (30 ml) Old Tom gin
1 oz (30 ml) sweet vermouth
2 dashes orange bitters
Orange rind, to garnish

### METHOD
Half-fill a small shaker with ice, add the liquid ingredients and stir with a bar spoon for about 30 seconds. Strain the drink into a chilled rocks glass.

To garnish, peel a strip of orange rind using a vegetable peeler. Fold to express its oils into the glass.

**RIGHT:** *The Ampersand*

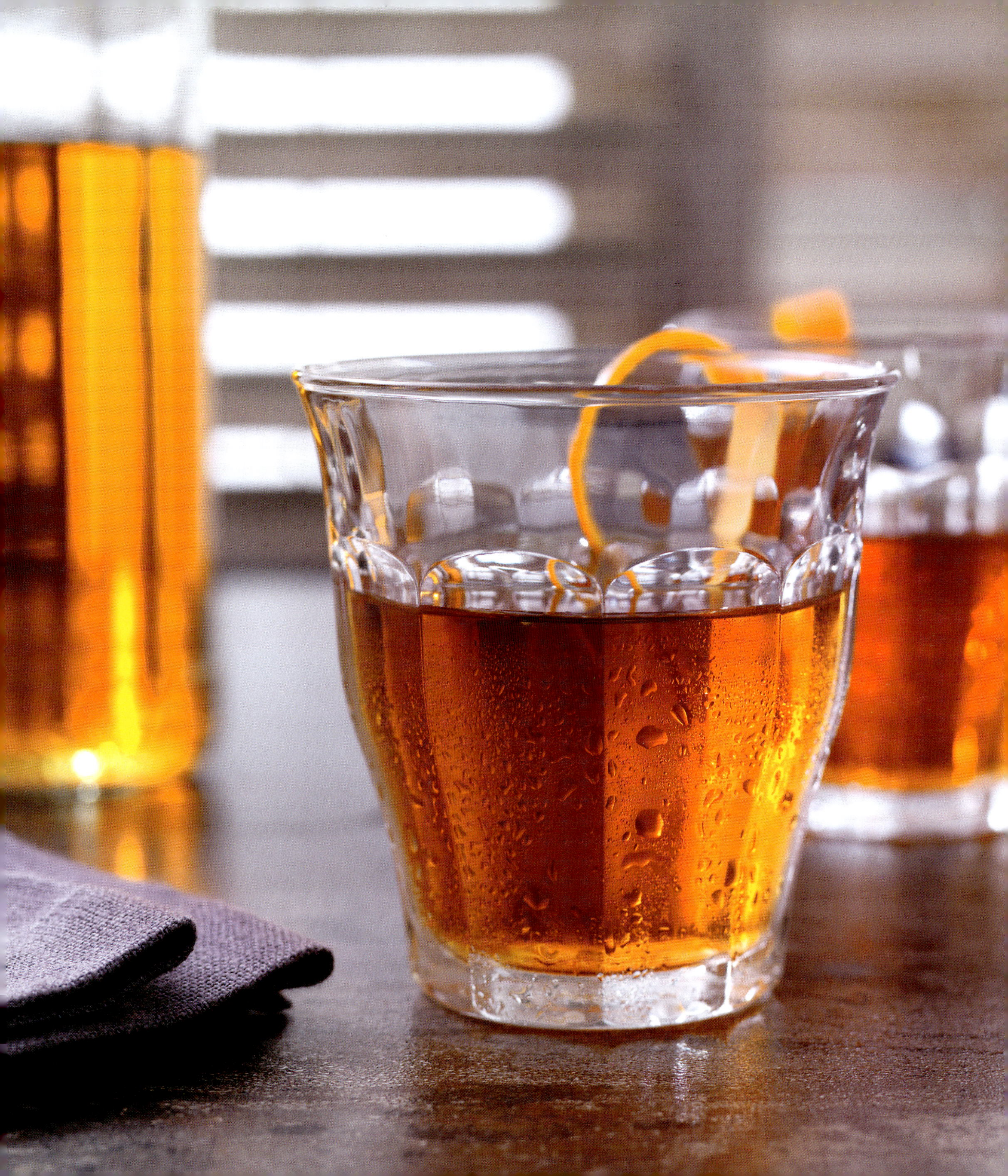

# PISCO BUCK

*Part of Copper & Kings' mission is to dispel the notion that brandy is only for cold weather consumption. The Pisco Buck is a great example of a tall, warm weather sipper created to cool and relax the drinker.*

### INGREDIENTS
2 oz (60 ml) immature brandy
¼ oz (7 ml) freshly-squeezed lime juice
3 drops Angostura bitters
Ginger ale
Lime rind, to garnish

### METHOD
Fill a highball glass with ice, add the brandy and the lime juice, and stir with a bar spoon for 10 seconds to chill. Add the bitters and the ginger ale and stir gently to blend.

To garnish, peel a thin strip of lime rind using a vegetable peeler and twist to express its oils into the glass.

# CORPSE REVIVER

*As one can easily infer from the name, the Corpse Reviver is known as a "hair of the dog" hangover cure. But since such "cures" rarely work, it can also be considered as a spirit lifter after a hard day's work. The apple brandy lends it a gentle charm.*

### INGREDIENTS
1 oz (30 ml) craft distilled brandy
1 oz (30 ml) unaged apple brandy
1 oz (30 ml) sweet vermouth

### METHOD
Half-fill a small shaker with ice, add all the ingredients and stir for about 10 seconds using a bar spoon. Strain through a julep strainer into a chilled coupe or rocks glass.

# VIEUX CARRÉ

*Named after New Orleans' French Quarter, the Vieux Carré's three-alcohol base leaves room for variation. For a softer note, back off on the rye or vermouth and boost the brandy. For more bite, boost the rye. Keep the alcohol bold, with 3 ounces (90 ml) per beverage.*

INGREDIENTS

1 oz (30 ml) aged brandy
1 oz (30 ml) rye whiskey (Rittenhouse Rye and Knob Creek Rye are great)
1 oz (30 ml) sweet vermouth (Carpano Antica Formula is extraordinary)
½ oz (15 ml) Benedictine
2 dashes Angostura bitters
2 dashes Peychaud's bitters
1 cherry and lemon rind, to garnish

METHOD

Half-fill a small shaker with ice, add the liquid ingredients and stir with a bar spoon for about 10 seconds. Using a julep strainer, strain into an ice-filled rocks glass.

To garnish, peel a thin strip of lemon rind using a vegetable peeler, twist to express its oils into the glass and place in the glass, along with a cherry.

# ROUND ROBIN

*If your palate isn't familiar with absinthe, this is a gentle introduction. It's a bold, herbaceous spirit that pushes notes of anise and fennel directly to the fore. The Round Robin is a great way to try it, since good brandy and frothed egg white soften the blow so nicely.*

INGREDIENTS

1 oz (30 ml) brandy
1 oz (30 ml) absinthe
1 tsp (5 ml) simple syrup (1:1 water to sugar)
1 large egg white
2 dashes orange bitters
Orange rind, to garnish

METHOD

Half-fill a small shaker with ice, add all liquid ingredients, cover and shake vigorously for 30 seconds to ensure the egg white is fully frothed. Strain the contents into a chilled coupe or rocks glass using a Hawthorne strainer.

To garnish, peel a thin strip of orange rind using a vegetable peeler and twist to express its oils into the glass.

**RIGHT:** *Round Robin*

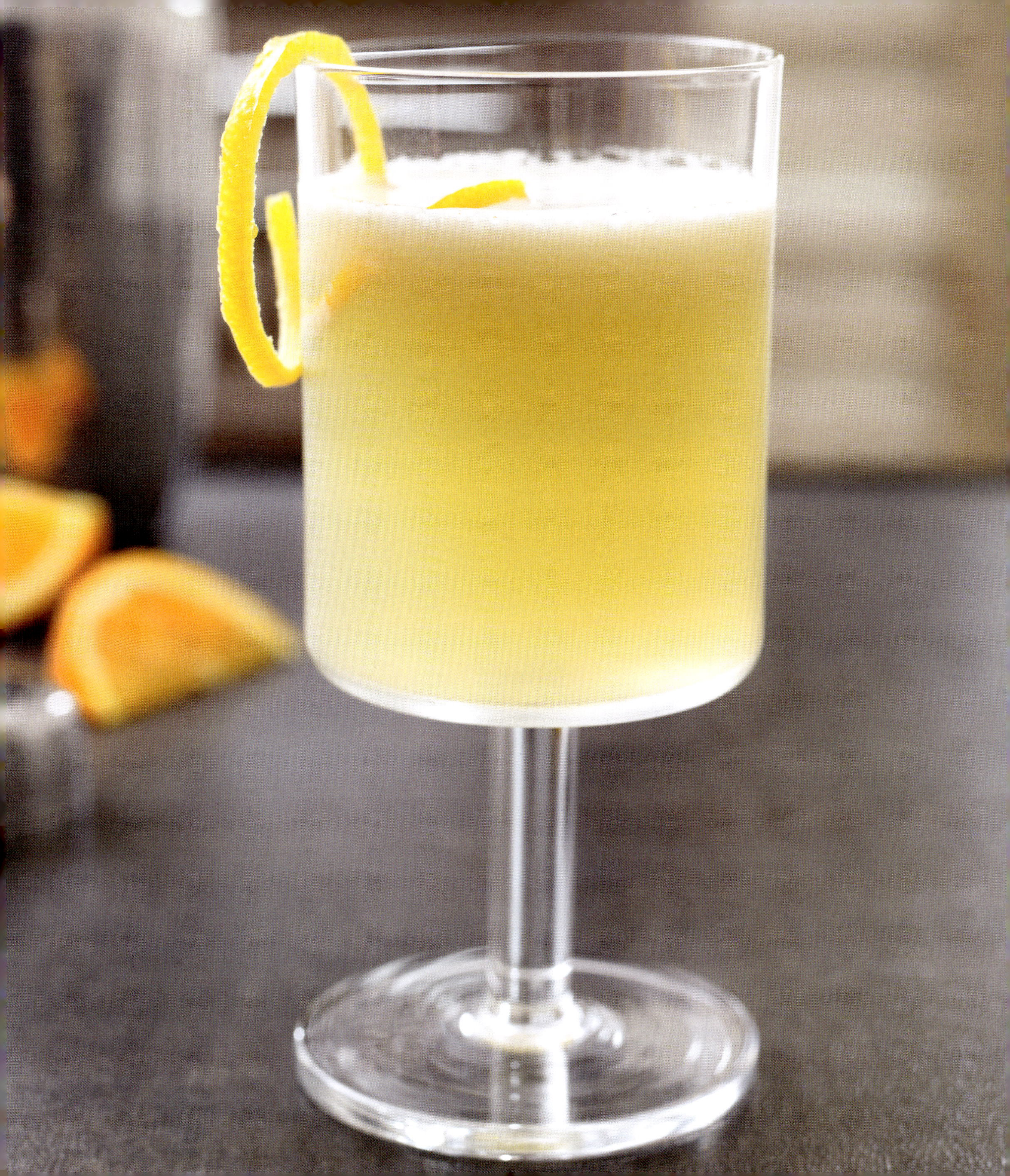

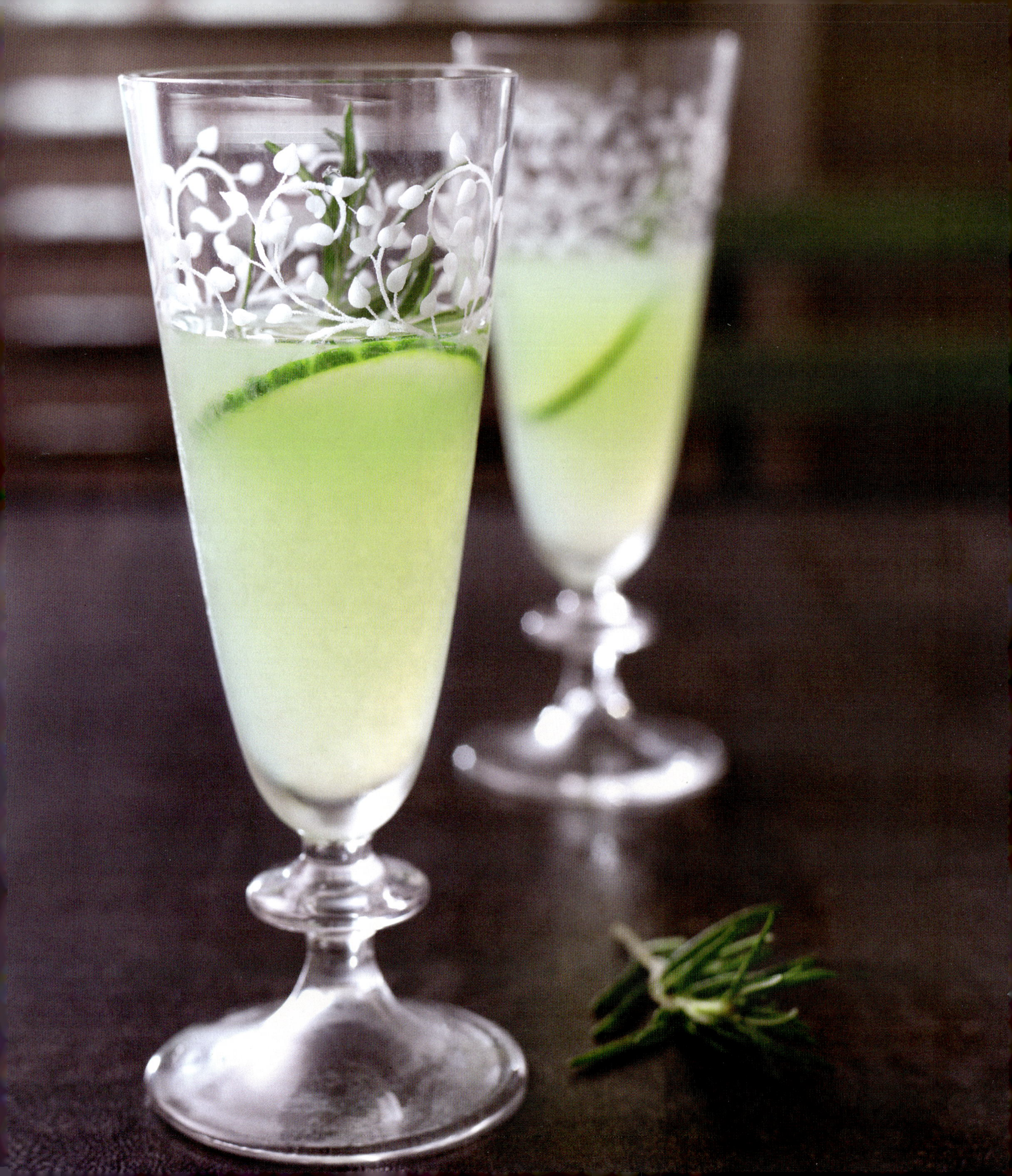

## THE IMPRESSIONIST

*This cocktail was created for Copper & Kings (see pages 126–129) by bartender Marie Zahn. The addition of fresh rosemary amplifies the already herbaceous absinthe, while the neutral cucumber scrubs off any rough edges in the mix.*

### INGREDIENTS
1½ oz (45 ml) white wine
1 oz (30 ml) absinthe
1 oz (30 ml) freshly-squeezed lime
2 sprigs fresh rosemary
4 thin cucumber slices

### METHOD
Half-fill a small shaker with ice, add all the liquid ingredients, one sprig of rosemary and two slices of cucumber. Cover and shake for about 5 seconds and pour into a Collins glass. Garnish with the reserved cucumber and rosemary.

**LEFT:** *The Impressionist*

## ST. CHARLES PUNCH

*Depending on which story you hear, this dark punch was named after either the St. Charles Hotel or the legendary St. Charles Avenue, both in New Orleans. Given the bawdy Mardi Gras parades that wind down St. Charles, this dignified drink almost seems out of place there.*

### INGREDIENTS
1 oz (30 ml) brandy or cognac
2 tsp (10 ml) simple syrup (1:1 sugar and water)
2 tsp (10 ml) freshly-squeezed lemon juice
2 oz (60 ml) ruby port
Fresh fruit, to garnish

### METHOD
Blend the lemon juice and simple syrup in a small shaker, then fill it about halfway with cracked ice. Follow with the port and brandy, cover and shake vigorously for about 5 seconds. Fill a Collins glass with ice and strain the drink into the glass using a Hawthorne strainer. Garnish with skewered fruit.

# INFUSIONS

*Of every potent potable discussed in this book, creating infusions is the easiest. Just add a flavor component to a base spirit for a prescribed length of commingling time, add sugar to sweeten — or not if you're using herbs — and bottle it. In practice, the effort is far more nuanced, and the devil is in the details, especially if distilling. For the past five centuries, infusions have been a reliable means by which farmers preserved excess produce using alcohol. The aim of the hobby distiller, however, is to produce intensely flavored booze for sipping, cocktails or food.*

*The possibilities of liquor infusions are endless; be they liqueurs or cordials; bitters, schnapps, flavored vodkas or spiced rums. Behind the hundreds available on the market, there are likely to be hundreds more on hobbyists' shelves.*

RIGHT: *Homemade infused vodka.*

If you were to ask me if I'd ever had the bad luck to miss my daily cocktail, I'd have to say that I doubt it; where certain things are concerned, I plan ahead.

— LUIS BUÑUEL, FILM DIRECTOR

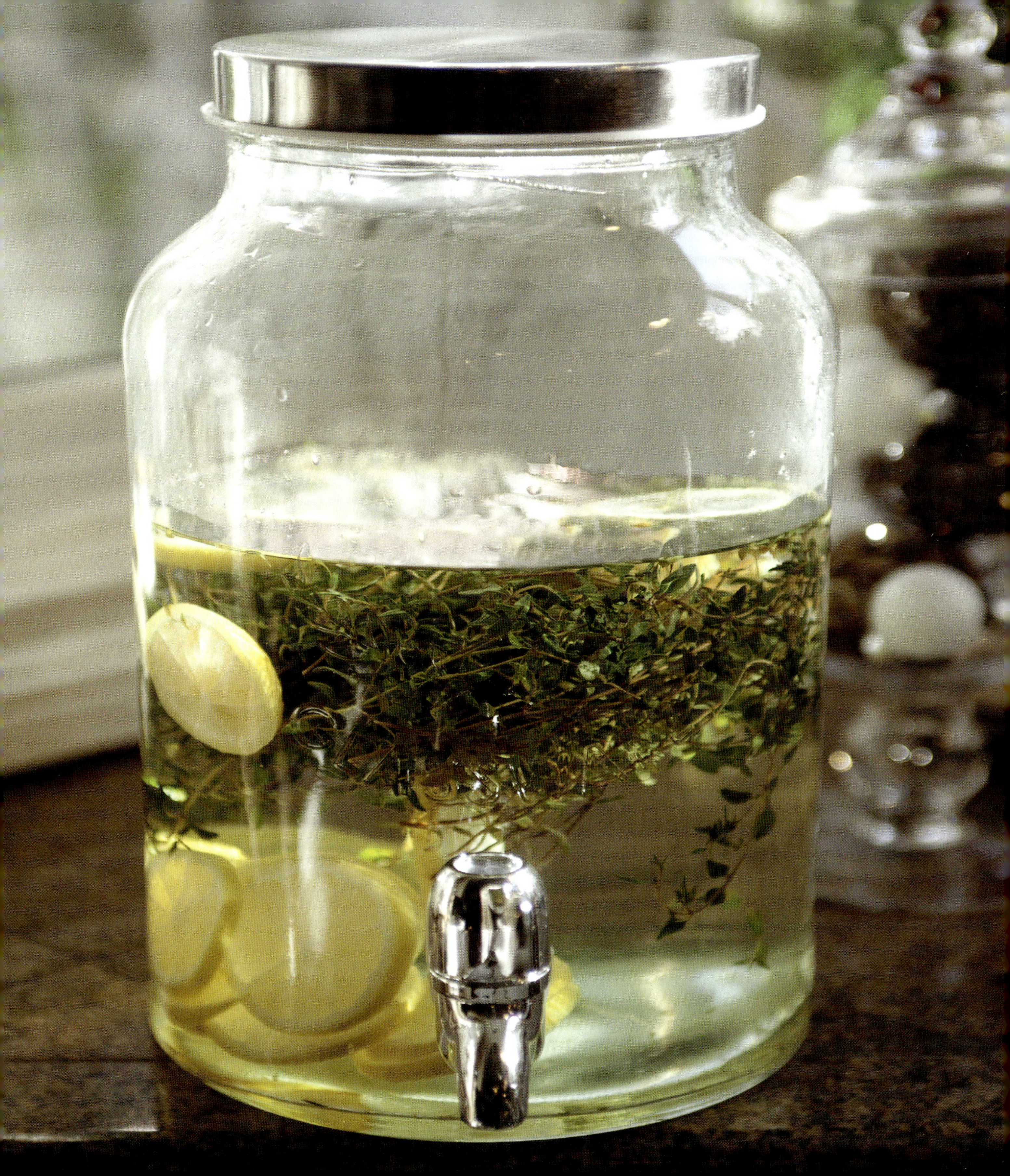

# A FEW DEFINITIONS

## Infusions

The simplest are passive infusions that combine fruit, spices and other flavorings with vodka, blanco tequila or white rum. They're left to rest, usually at room temperature, but sometimes in the refrigerator. Depending on the flavoring agent, transfer can be fast (dry spice flavors migrate quickly) or slow (fresh fruits and vegetables since they contain water). Usually such infusions are used as base spirits in cocktails.

One of the best-known infusions is spiced rum, whose prominent flavoring agents include vanilla, allspice berries, cloves, nutmeg, ginger and orange rind. Increasingly common are fresh pepper infusions: heat from jalapeño slices added to tequila penetrates quickly, turning a standard margarita into an exciting and memorable drink.

The still also is employed in infusions. Flavoring agents (such as fruits or spices) are steeped in a neutral spirit and distilled. It's that simple.

But even easier is taking a neutral spirit, watering it to your desired final proof and adding one of dozens of flavorings made from high-quality extracts. Even if you frown upon such shortcuts in your quest to make an infusion from scratch, these easy blends are good for the quick production of a few "homemade" options to shelve while learning the authentic way.

## Liqueur

This is made from a distilled spirit flavored with fruit, herbs, spices, flowers or nuts. While most are sweetened with sugar or sugar syrup, many are herbaceous and bitter — flavors increasingly sought out by skilled mixologists. Chartreuse (created four centuries ago), Cynar, Punt è Mes and Zucca are good examples of herbaceous and bitter options. Sweet options are endless: triple sec, crème de cassis, Chambord, chocolate, cream and so much more. The middle ground blends the two for results such as licorice-accented Galliano and the legendary anise and wormwood-spiked absinthe.

Fruit liqueurs are steeped, meaning the spirit and fruit are in direct contact in the same vessel to maximize extraction. Not all fruit liqueur recipes require heating the spirit solution, though many do. Heat does extend a liqueur's shelf life, and can alter the delicate flavors you're trying to infuse by making it taste cooked instead of fresh. Fruit liqueurs also are regularly distilled to refine their flavors.

Plant liqueurs are flavored through percolation or distillation. In percolation, plant material is held in a perforated basket above a boiling spirit. The boiling spirit ascends through a pipe to the top of the percolator and onto the plant material. It then flows through the perforated basket and back down into the percolator where the cycle recurs.

In distillation, the plant material is steeped in a heated spirit and allowed to macerate. Once flavor extraction is completed, the whole is distilled. Purists argue that nothing beats steeping for maximum aroma and flavor extraction. The makers of Big O Ginger liqueur (see page 146 for profile), add 22 pounds (10 kg) of fresh chopped ginger to 100 gallons (380 l) of blended grape brandy and syrup. In making their 135 proof absinthe, distillers at Copper & Kings American Brandy (see featured profile in the Brandy chapter, page 128), steep 50 pounds (22 kg) of dried botanicals for 24 hours in 500 gallons (1,900 l) of stripped muscat grape wine before distilling the mixture.

## Bitters

These high-proof infusions are flavored with a blend of pungent and mild plant extracts. With just a few drops, bitters add spice and contrast to sweet notes in most cocktails. Bittering agents include gentian, quassia and wormwood, among many others. Milder flavoring agents in bitters include orange, grapefruit, celery, cardamom and ginger. These are designed to amplify similar flavors born of other cocktail ingredients such as juices.

## Schnapps

Traditionally, schnapps are made from fruit fermented (for a few weeks to a few months) with sugar and a base spirit, and then distilled. Usually the base spirit is brandy, which makes sense, given its fruit origin. Quicker versions combine high-quality fruit juice and

**ABOVE:** *Plum brandy, or "slivovitz", in Orebic, Croatia, where damsons are fermented with sugar and a base spirit before being distilled to produce schnapps.*

100 proof spirits, which are then distilled without fermentation. Well-made schnapps are designed for relaxed sipping, never shooting — but to each his own.

## A Word About Potency

While many infusions are potent at 80 proof, many are half that ABV and even lower. High-proof infusions such as absinthe (130 proof and up) or Orange Curacao (80 proof) deliver incredible flavors and aromas, but managing the sheer supply of alcohol these supply in a single drink takes finesse on the part of the mixologist. By comparison, lower-proof infusions such as Amaretto Disaronno (52 proof), Aperol (22 proof) and Campari (28.5 proof) are equally flavor-forward, yet they don't deliver such a powerful punch of alcohol, so are often mixed with other spirits or soft drinks.

Bottom line: Suit yourself and pour according to your palate's desires. With most infusions, a little goes a long way, so start with small amounts and add more if you please.

## Author's Note

The making of passive alcoholic infusions has occasionally pitted bartenders against the Department of Alcoholic Beverage Control in the state of California. Authorities there say taking spirits, infusing them with a flavoring agent, such as fruit, and then allowing them to mature over time, makes a rectified spirit. Rectification without a special license is illegal there.

The *Encyclopaedia Britannica*, however, defines rectification as "… the process of purifying alcohol by repeatedly or fractionally distilling it to remove

**ABOVE:** *Soft fruits such as berries will release their flavors into the spirit quickly, and work particularly well with vodka.*

water and undesirable compounds." Rectifying means changing, and, chemically speaking, neutral spirits are not changed by the introduction of fruit. Such simple infusions are blends that are no different from fruit or herbs muddled and mixed into a cocktail. With all due respect to California's DABC, you're overreaching here.

## FRUIT-INFUSED VODKA

The work of distilling is easily forgotten when you get to play with the finished product. And few alcoholic drink recipes are as simple to make as fruit infusions. Given vodka's neutrality, it's not surprising that it's most commonly used for infusions, but plenty of great infusions are made with bourbon, brandy, gin, tequila and rum.

The work of infusing fruits is simple: Cut up fresh fruit, cover it with spirit, let it rest a few days and begin tasting it to see when you think the flavor transfer is at its peak. Strain out the spent fruit, bottle the liquid and you're finished.

As you become more experienced with infusions (and don't forget to make tasting notes on what worked well and what didn't), learn to experiment with spices and herbs, as they add depth and complexity to the fruits' sweetness.

### Proof Choice

Whatever spirits choice you make, 80 proof will suffice, while 100 proof will extract more fruit flavor. If the final result is too intensely alcoholic for your palate, proof it down with a thin simple syrup (1:1 — or even 1:0.75 — water to cane sugar). It's better to err on the side of fruit intensity, so consider 100 proof, and modestly priced spirits at that.

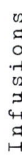

**ABOVE:** *Rinds and skins make for punchy vodka infusions.*

**ABOVE:** *Quince schnapps is a wintry favorite.*

### Solids, Liquids, Time

A good rule of thumb is 2 cups (250 g) of chopped fruit to 2–3 cups (500–700 ml) of spirits — whatever is required to fully cover the fruit and delay spoiling. Infuse that mixture in a tightly sealed jar for three days and taste it. If it's not strong enough, wait two more days and taste again.

Soft fruits, such as strawberries or blueberries, will release their flavors more quickly than firmer fruits, like pineapple or cherries. To boost flavor transfer, one bartender I spoke with takes fresh cherries and steams them until tender before adding them to his vodka. Another I know grills fresh pineapple slices to add smoky notes to his infusion.

Fruits with high water content, such as grapefruit and orange, aren't ideal when used whole. Their real flavor punch comes in their rinds. When peeling citrus fruits, make shallow strokes to avoid digging into the white pith. Gin, tequila and rum are particularly friendly to lemon and orange rind infusions.

Once finished and strained, infusions will last a long time — likely longer than it takes for you to drink them. If you're at all nervous about spoilage — which you shouldn't be if you chose 100 proof liquor — refrigerate them.

Since these aren't liqueurs, drinking them straight isn't recommended because they're just not refined. Like bitters, fruit infusions are best used to flavor cocktails or foods. Better yet, they make great and inexpensive gifts for friends.

# BASIC FRUIT-INFUSED VODKA

*A basic fruit-infused vodka is described here but you can add any of the following ingredients for an optional flavor boost and to add real pizzazz to your infusion: whole cloves, cinnamon sticks, minced fresh ginger (peeled), vanilla beans and citrus.*

## INGREDIENTS
2 cups (250 g) chopped ripe fruit
2–3 cups (500–700 ml) vodka
 (80 proof or 100 proof)

## METHOD
Place the fruit and alcohol in a 1 quart (1 l) sealable jar and let it sit for at least three days before tasting. Strain through cheesecloth and a fine sieve into a second vessel; take care to press only lightly on the fruit. Using a funnel, transfer strained infusion to a bottle. Store in a cool place and out of direct sunlight.

**RIGHT:** *Traditional fruit-infused vodka.*

HOME INFUSED
LEMONGRASS
VODKA
Home Infused
Lemon Verbena
Vodka.
Vodka

# WORDS OF WISDOM

## KATHY, BILL AND CHRIS FOSTER

### OWNERS, BIG O GINGER LIQUEUR

Maplewood, Missouri
www.drinkthebigo.com

A taste of well-made limoncello sipped on a visit to London years ago made Bill Foster and Kathy Kuper eager to replicate the liqueur at home. And despite their first attempts winding up in the trash, Bill, who was handy in the kitchen, kept after his goal, albeit with a course change to a ginger liqueur. After about a year-and-a-half, they were so happy with the results that they started giving it as gifts to friends and family. Eventually, the endless encouragement they received from those recipients convinced them to begin making it professionally, which led to them launching Big O Ginger Liqueur in 2009 with their son, Chris Foster.

**There are a lot of liqueurs on the market, so what makes yours unique?**

**Bill:** Ours is a lower-proof sipping liqueur, which is why it makes better cocktails. Kathy didn't want it to be hot on the palate, so we kept the alcohol low. Also, how we handle the ginger and spices we put in gives it a unique flavor profile.

We start with grape vodka made from Missouri wine, which makes the texture very smooth and round and the flavor quite crisp. We macerate by using fresh ginger, which we cut ourselves, two kinds of unrefined sugar and a little citrus. We cook it and put it into a tank with the spirits we use. One's a vodka and the other is the brandy, which is 142 proof, and we let it macerate one month in stainless steel.

**Chris:** If it steeps too long, you get a bitter taste, and we want it to have a nice finish: smooth on the back of the tongue. In a liqueur you have the chance to push flavor nuances and take on bold flavors that you're not going to get out of distillation.

**I'm guessing you use a lot of ginger.**

We hand chop 22 pounds (9 kg) of fresh ginger for every 100 gallons (378 l) of the spirits and syrup mixture. It gives it a really nice, spicy flavor without being overwhelming.

**And why ginger?**

**Bill:** When I was developing the recipe, I really tapped into the flavors I liked, and I was inspired by Asian cuisines. If I were to tell others who want to do their own liqueur, I would say explore flavor combinations that you already find pleasing to drink. And don't be surprised when serendipity throws you a curve ball that turns out to be a fantastic flavor. That happened to us one time when I was playing with the recipe and I was out of vodka. We had some relatively inexpensive brandy, and Kathy said, "Oh throw that in." And it was the perfect rounding for it.

**Chris:** But when serendipity strikes, be ready to write it down. Anytime you make a change, do a controlled experiment. Change one thing at a time only so you can get some consistency in the end.

**How did you scale the recipe for large production?**

**Kathy:** We took the science route and talked to a chemist and chemical engineers. But if you have a love of cooking and experimenting, go with your gut.

**Was it difficult to move from gift makers to business people?**

**Kathy:** We got our license in 2009, but it took us two years from that time to get all the proper permitting and get Big O on the market by Father's Day in 2011. We were surprised it took that long, but we have a friend who'd worked at Brown-Forman for 30 years, and she was flabbergasted that it happened that quickly.

**Bill:** Originally we were looking into opening a distillery, but we figured out we didn't need to. We went to High Plains Distillery in Atchison, Kansas, to talk about it, and they said we could spend $300,000 to build our own and still not have any product on the market. Or, they could make it for us, get the product to market right away, without spending so much money. That turned out to be the right thing to do. Since then, we have moved to a distillery in St. Genevieve, Missouri.

**ABOVE:** *Bottle of Big O.*

**What is your annual case volume?**

**Bill:** 10,000 bottles a year, and those go into six-pack cases. It turned out that it was easier to get stores to take our product if we had cases of six rather than cases of 12.

**Kathy:** That's partly because distributors have an upcharge if you don't buy the whole case, so it was easier for the customer not to pay the upcharge.

**Did you know anything at all about the liqueur business before you got into it?**

**Kathy:** Oh, no, we had no idea what we were doing. Bill was a college professor for 30 years and I was a trained photojournalist. We were fish out of water. Chris has a business degree and was the one who put together a business plan.

**Chris:** We asked a lot of people a lot of questions, but that's tricky too. I've heard at a distillers' class that if you ask six distillers one question, you'll get eight answers. But I do know that I'd pay a lot of money for the notebook they gave me at the Distilled Spirits Epicenter class. I learned a lot there.

# ORANGE SCHNAPPS

*Since orange pairs so well with spirits and functions in so many cocktails, it makes sense to have a bottle of orange schnapps handy. You can blend it with other schnapps in order to enhance flavor, or drink it on its own as an appetizer. It also works well in a main dish of duck, and as an ingredient in chocolate desserts.*

## INGREDIENTS
2 medium oranges
9 oz (260 ml) vodka (80 proof)
1 oz (30 ml) simple syrup (1:1 sugar to water), optional

## METHOD
Using a vegetable peeler, remove orange rind without white pith. In an oven set at 150°F (65°C), or in a dehydrator, dry peels for about an hour.

Place dried rind and vodka into a small, sealable jar. Store in a cool, dark place and shake lightly daily. After three days, taste the schnapps. If you want more orange flavor, let it age for three more days, then check again. Just beware that steeping too long can make it bitter.

When the desired orange flavor is achieved, remove the rind, stir in simple syrup if desired, and strain through a fine sieve into a bottle. Seal or cork the bottle and store in a cool, dark place when not imbibing.

# RASPBERRY SCHNAPPS

*A weaker spirit thanks to the low sugar content in raspberries, this delightful schnapps is great for sipping as is, or as a refreshing accompaniment to fizz on a summer's day. You also can add heavy simple syrup (2:1 sugar to water) to this recipe to make it suitable for food use.*

## INGREDIENTS
1 pint (275 g) fresh, ripe raspberries
12 oz (350 ml) vodka (80 proof)
1½ oz (45 ml) simple syrup (1:1 sugar to water), optional

## METHOD
Rinse raspberries and let them dry on paper towels. Add berries to a 1 quart (1 l) sealable jar, followed by vodka. Store in a cool, dark place and allow to steep for three months. Shake jar lightly weekly.

After steeping, pour infusion through a coarse strainer into a bowl. Using a funnel fitted with a fine sieve or three layers of cheesecloth, pour the bowl's contents into a storage bottle. If berry particles settle out in the bottom of the bottle, consider filtering them through a coffee filter for the clearest result. (Fewer solids in your schnapps means better shelf life.) Add simple syrup if desired.

# ABSINTHE

> When Henri-Louis Pernod first produced absinthe commercially in France in 1805, it is unlikely that he could have predicted that it would become the stuff of legend, far greater than even the most delectable distilled grape wine ever deserved. But it has, despite long bans in some countries, made a comeback and regained its popularity.

Parisians couldn't get enough of the super-potent anise and wormwood spirit in the 19th century, drinking it sometimes before a workday and nearly always afterward, in cafés and social settings. How it became viewed as simultaneously devilish and irresistibly delicious isn't completely clear, but what's certain is the habit of overindulgence was at the core of both.

Commoners claimed absinthe made them hallucinate, and artists, poets and writers credited it with inspiration — it is said that Van Gogh cut off his ear when under its spell. Yet despite such popular, claims, people were not put off, and many sought to sip "the green fairy" to sample its effects. In their book *Distilled*, Joel Harrison and Neil Ridley called it the LSD of the day, adding that Oscar Wilde claimed absinthe's aftereffects left him feeling as though tulips were on his legs. However, as Sarah Hepola, editor at Salon.com has said, "Much is said about absinthe; very little of that is true."

Predictably, those on the sidelines watching others consume absinthe weren't imagining the sight of lots of extremely drunk people struggling to shake its grip on their livers and minds. Historians compare absinthe's heyday to the raucous and raunchy decades of England's Gin Craze in the 18th century. And with its proof ranging between 110 and 140, it's no surprise that people got a little too loose. Overconsumption of beer, wine or whiskey will yield the same dazed results, yet somehow this richly herbaceous and spicy liquor earned the reputation of being both mind slayer and soul commander. Even after being banned in many countries for much of the 20th century, absinthe returned to the bar with its sinister reputation intact.

About the only thing absinthe truly dominates is one's palate, delivering a complex array of flavor and texture. Anise leads, followed by floral notes buttressed by peppery spice and herbs, such as mint and hyssop. The distiller's true skills are showcased in the application of its bittering agent, wormwood, to counter the spirit's livelier flavors. While just a little wormwood goes a long way — it's horrid consumed by itself — absinthe wouldn't be the same without it.

If you dare, sip a ½ ounce (15 ml) or less straight (well-made absinthe should sting a little upon entry, but not burn on exit) and consider it while rolling it around your mouth. The spirit's grape base will emerge between more delicate flavors. Sniff it and taste it like you would any spirit, and you'll find while anise never steps back, it never bullies. It's like a protective sibling accompanying all other flavors and aromas as if charged with introducing each personally.

Traditionally, absinthe is poured over a sugar cube resting on a perforated spoon placed atop a cocktail glass of water. As the absinthe contacts the water, it turns it cloudy white as its essential oils precipitate out of the alcoholic solution. This reaction is called the *louche* and it's a sight to watch, especially when the liquor is set aflame and caramelizes the sugar.

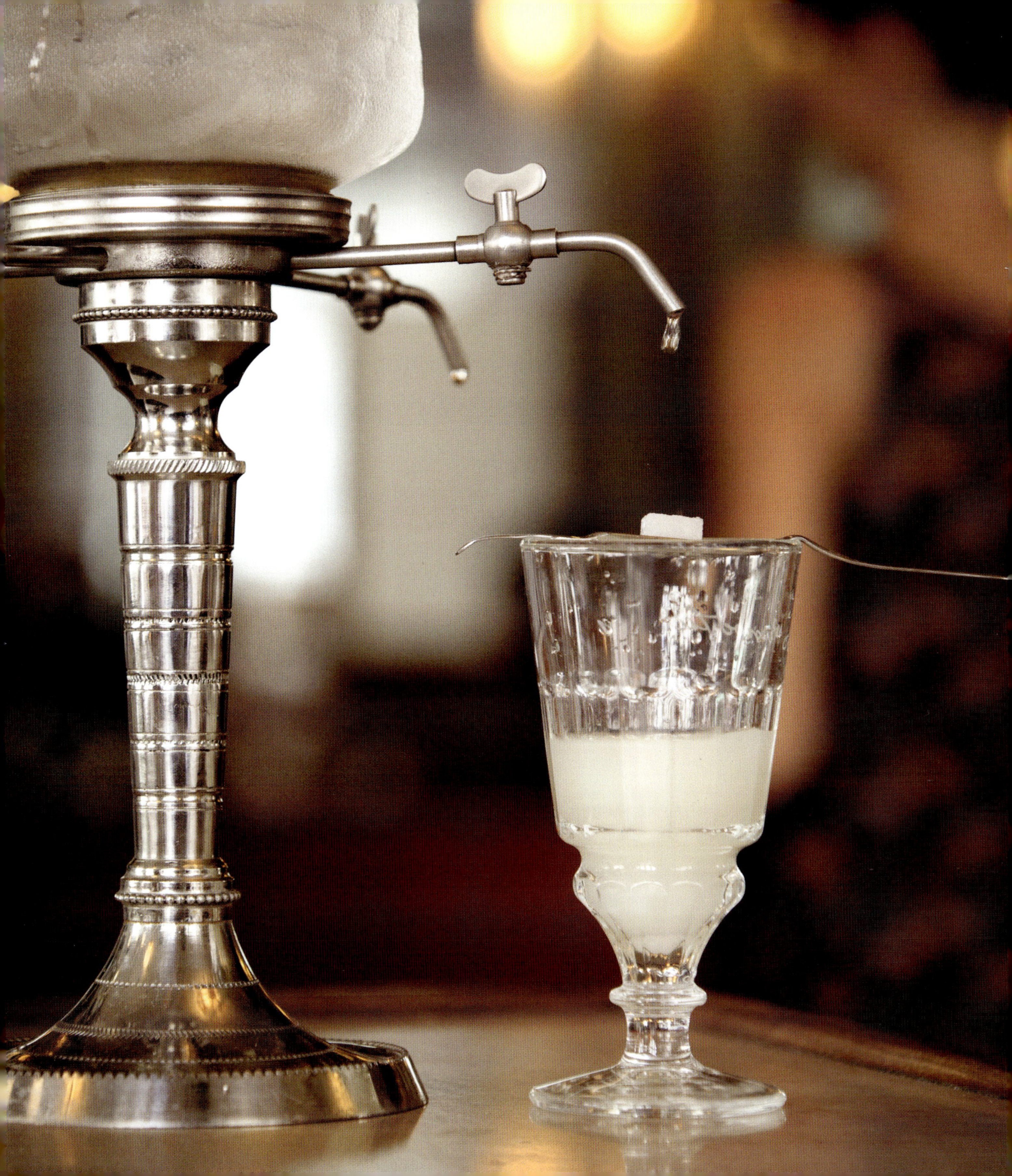

# WORDS OF WISDOM

## JUSTIN KING
### MASTER DISTILLER, OLE SMOKY MOONSHINE
Gatlinburg, Tennessee
www.olesmoky.com

As a product of East Tennessee, it's not surprising Justin King grew up around moonshining. It was both a family passion and an income stream for his great grandmother, who had multiple mouths to feed on a farmer's wife's budget.

King carried on the tradition as a hobby, creating highly desired Christmas gifts that, not surprisingly, no one returned. Especially not family friend Joe Baker, who asked King to work for him as a distiller at Ole Smoky Moonshine when he founded it in 2010. In the years since, King has helped create a slew of award-winning moonshines and whiskeys.

**How different is working as a professional distiller from being a hobbyist?**

We pay our taxes and we don't have to look over our shoulders anymore! This is a dream job because I get to play with 800 gallon (3,000 l) professional pot stills every day. But the truth is it's no different from what I did with a small still. It's surprisingly similar because Ole Smoky is also a craft distillery; nothing is automated.

**Talk about some of your products.**

Our moonshine is a corn whiskey made from 80 percent corn and a 20 percent mixture of rye, barley and wheat. We run that in an 800 gallon (3,000 l) pot still, and it comes off at about 80 proof on the first run. We take that 120 gallons (450 l) of alcohol and run it again to get 155 proof and 90 gallons (340 l) of alcohol. We then water that down to 100 proof. We also do a charred moonshine whiskey, and it spends six months in the barrel.

**What's "charred moonshine"?**

When I made moonshine illegally, if we could get a used barrel to put it in and give it color and taste, we called it charred moonshine. Since this is still young, we call it a harsh barreled product; it basically has some of its edges worn down by the barrel.

**I'm really interested in your Blue Flame Moonshine, mostly because it's 128 proof.**

It's our new sugar product. When we make corn whiskey, we take some of the spent mash and add white cane sugar and some water back to it. We get the gravity level right, and ferment that to about 12 percent alcohol. We distill it to 128 proof using a pot still with a thumper on it.

🍶 **And why the name Blue Flame?**

Well, when the light is low, you can light it and it will burn a really pretty blue flame.

🍶 **Of all the brand's products, what's the most popular?**

Our Apple Pie Moonshine is by far. At 40 proof, it's really easy to drink. Just take the lid off and start sipping.

🍶 **Talk a little bit about your family's heritage as moonshiners.**

Well, when you're like my great grandmother and you have seven kids, you had to make some choices. She could take a bushel of corn and get a dollar out of it as is, or take that same bushel and make liquor and sell it for $20. That's how you fed your kids. Her family would make it in the fireplace inside the house.

We did a true corn whiskey. People around here would make their malt by putting corn in a burlap sack and setting it in the creek for a day. The next day they would take it out, set it in the sun and let it sprout. Once it sprouted an inch, they dried it out using a wood fire — if you didn't have a fan — or they put it in a kiln. After that, they ground it. When Joe created Ole Smoky Moonshine, that's what he was looking for: an authenticity that people would appreciate.

🍶 **Are you surprised that moonshining, which you did for fun, has become so popular?**

In a way I'm not surprised. There is a mystique to it for sure, and people are getting a taste of something made the way it was 100 years ago. Products like our Blue Flame or our corn whiskey are things you have never been able to go out and get legally. Now they can go to the package store and get the real thing legally.

**ABOVE:** *Ole Smoky moonshine on sale at the distillery.*

🍶 **What would you advise hobby distillers to do to improve their craft?**

Take a distilling class. In North America there are so many classes. I'd also tell people to visit distilleries that do tours to see the process in action. We do tours as does Davy Crockett's (Tennessee Whiskey) near us. It gives you a chance to understand what good mash is supposed to look and smell like. You can see and smell and taste everything we do by visiting one place.

🍶 **Do you believe home distilling should be legalized?**

Well, yes and no. I can understand the government's reason: the tax money isn't there. When we are paying $13.50 U.S. per proof gallon, and somebody makes 10 gallons at home, that's $135 of tax money not paid to the government. But you can ferment wine and beer at home tax-free and that's legal, and that's not really fair. So I don't guess I've really decided.

# GLOSSARY

**ABV.** Short for "alcohol by volume," which is measured as a percentage of alcohol suspended in a liquid solution. For example, a 100 gallon wash with an ABV of 8 percent contains 8 gallons of alcohol.

**Aeration.** The process of circulating air into a liquid through modest agitation. In fermentation, aeration provides yeast with the oxygen required to multiply in a wash or mash.

**Aging.** Storage of distilled spirits in wood barrels to enhance and alter their flavor and texture. A hybrid form of aging is adding wood chips or pieces of barrel staves to distillate stored in nonwood containers.

**Airlock.** A one-way valve that, during fermentation, allows $CO^2$ gas to escape while blocking outside air from entering the fermenter.

**Alcoholmeter.** An instrument calibrated to read a liquid's density based on its percentage of alcohol. This is used to measure alcohol content in a finished spirit.

**Alembic still.** The oldest known spirits distiller, its bulbous boiler base rises to a narrow cap that leads to a descending condensing arm where vapors begin to condense and run into a collection vessel. It is the forerunner of the pot still in use today.

**Alpha amylase.** An enzyme used to break down long-chain starches into sugars (glucose) that are easily accessed and fermented by yeast. This process is known as liquefaction.

**Angels' share.** Alcohol that evaporates during aging.

**Antifoaming agent.** An additive used to reduce foaming during fermentation or distilling.

**Backset.** Liquid remaining at the bottom of a still after distillation, and which is added to a new wash or mash to encourage fermentation and produce flavor consistency between batches.

**Beer.** A low-alcohol liquid produced by fermenting cereal grains.

**Beta amylase.** An enzyme used to break small chains of sugar into smaller pieces (maltose). The process is known as saccharification.

**Brix.** A measurement of the amount of sugar dissolved in a liquid solution. In distilling, knowing brix allows one to calculate the potential amount of alcohol one could produce through fermentation. Brix is most widely used in winemaking, hence its importance to brandy distillers. Brix can be measured using a hydrometer.

**Charring.** The act of burning the inside of a wood barrel to impart character, color and sweetness to aged spirits; also referred to as "toasting."

**Clearing agent.** An additive placed into a wash to accelerate the movement of dead yeast to the bottom of the fermenter; also referred to as a finishing agent.

**Congeners.** Substances produced in the fermentation of alcohol that can improve or degrade the flavor of distilled spirits. Among them are acetone, acetaldehyde, methanol, esters and aldehydes.

**Cooperage.** The place where barrels and casks are made, and also a term used for finished barrels and casks.

**Cuts.** Precise points during a distilling run at which distillate is collected at specific temperatures and alcohol proofs. SEE ALSO HEADS, HEARTS AND TAILS.

**Distillation.** Purifying a liquid by boiling it and condensing its vapor.

**Doubler.** SEE THUMPER/THUMP KEG.

**Equipment cleaners.** Chemicals (and tools) designed to remove buildup of by-products in distilling and fermenting equipment.

**Equipment sanitizers.** Chemicals used to remove contaminates from fermentation and distillation equipment. Sanitizers range from simple bleach to Star San, made specifically for alcoholic beverage production.

**Esters.** In distillation, esters are fragrant organic compounds formed by reactions between acids and alcohols as alcohol is boiled off a mash or wash.

**Ethanol.** A potable and therefore sought-after alcohol in distilling.

**Excise tax.** Taxes paid by the purchaser of specific goods, in this case liquor.

**Fermentation.** The process in which yeast converts sugar into $CO^2$ and alcohol.

**Fermenter.** A container for mixing and holding ingredients during fermentation. For home distillers, these commonly are food grade plastic buckets purchased from wine or beer making supply stores, or a glass "carboy," which is a giant jug with a sealable mouth that can be fitted with an airlock valve.

**Filtering.** SEE POLISHING.

**Finishing agent.** SEE CLEARING AGENT.

**Finishing run.** The distillation of the low wines produced in a stripping run. On this second distillation run (sometimes referred to as the spirit run), heads, hearts and tails are separated to make the final spirit.

**First distillation.** The initial distillation run of a wash or mash, also called the stripping run. Mostly water and some harsher alcohols are removed during this run.

**Flavoring (verb).** The addition of specific flavors such as fruits, herbs, botanicals or nut extracts to distillate. Flavoring can be done during or after distillation.

**Foreshots.** The first spirits (roughly 5 percent or less of the total distillate) to emerge from the still during distillation at about 176°F (80°C). Since these contain nonpotable chemicals such as methanol, acetone and aldehyde, foreshots are always discarded.

**Fractional distilling.** SEE REFLUX DISTILLING.

**Fusel oils.** By-products of alcohol distillation that include amyl alcohols, propanol and butanol. They are slightly viscous and malodorous, and appear in the tails cut of a distillation run.

**Grain-neutral spirit (GNS).** A spirit made from fermented grain and distilled to 190 proof. It is colorless, odorless and largely neutral in flavor. Grain-neutral spirits are consumed as vodka, redistilled with botanicals to make gin, diluted to create liqueurs, and blended with other spirits.

**Gravity.** A measurement of the amount of alcohol within a mash, wash or distilled spirit.

**Gravity, final.** A measurement of any unfermentable substances remaining in a wash or mash after fermentation. Final gravity is compared to original gravity to determine how much alcohol was created in fermentation. The number representing final gravity will be below the original gravity or the wash/ferment, but it will be above specific gravity.

**Gravity, original.** A measurement of the sugar content of a mash or wash to determine its potential alcohol. This pre-fermentation measurement is taken using a hydrometer.

**Gravity, specific.** A measurement of the density of a sugar dissolved in water relative to the density of water, which registers 1.0 on a hydrometer.

**Heads.** All alcohols coming off the still at 171°F (77°C) or lower. Since heads can be poisonous and produce unpleasant aromas and flavors, nearly all are discarded. However, some distillers collect some small amounts to flavor their final spirits.

**Hearts.** The portion of the distillate containing the greatest amount of desirable ethanol. This is collected during the middle of a distillation run within a temperature range of 172°F (78°C) and 203°F (95°C).

**Hydrometer.** An instrument for measuring the density of a liquid relative to pure water. In distilling, a hydrometer is used before fermentation to calculate the potential alcohol in a solution by measuring the water's density relative to the percentage of sucrose suspended within it. It is used again after fermentation to recalculate potential alcohol.

**Infusion.** The addition of flavor to spirits during distillation by suspending items such as botanicals or herbs in the vapor path inside the still. Infusion is different from steeping, wherein flavoring components are blended with stripped distillate.

**Krausen.** The layer of foam that forms atop of a wash or mash while fermenting.

**Lag phase.** The time period in which yeast propagates in a wash or mash.

**Lees.** Crushed fruit and yeast that accumulate during fermentation. Some brandy makers distill their mash with the lees, while others filter out only the liquid.

**Liquefaction.** The result of adding alpha amylase to a hot gelatinized mash. The process breaks down long molecular starch chains to smaller chains known as dextrins.

**Low wines.** The alcoholic beverage produced during the first distillation or stripping run.

**Maceration.** The addition of solids to low wines or spirits in order to extract their flavors. Some gin makers add botanicals to low wines, while liqueur makers macerate fruit and spices in vodka or other spirits.

**Malted barley.** Barley that is partially germinated, dried and then ground to grist.

**Mash (noun).** A mix of crushed or cracked cereal grains and hot water.

**Mash, mashing (verbs).** The act of steeping crushed or cracked cereal grains in hot water to gelatinize their starches.

**Mash bill.** The variable proportion of grains used in a mash for distilling. The mash bill also provides drinkers with clues about a grain whiskey's flavors.

**Maximum alcohol tolerance.** The percentage of alcohol that yeast can tolerate and continue to ferment sugars within a wash or a mash.

**Methanol.** A poisonous, nonpotable alcohol produced in the distillation of fermented beverages. It boils and emerges from a still at 148°F (64°C) and is commonly discarded in the foreshots and heads cuts.

**Moonshine.** Potable spirits distilled illegally.

**Packing.** Material, such as copper mesh or ceramic rings, that are suspended in a column still to promote refluxing.

**Parrot.** A holding vessel located between a still's condenser and its collection container. It catches distillate for the purpose of measuring its alcohol content with an alcoholmeter. The parrot allows alcohol content readings to be made during the entire distillation run.

**Pitching yeast.** The act of adding yeast to a mash or wash and blending it with the liquid.

**Polishing.** The use of filtration, such as activated carbon, to remove congeners remaining in the spirit after distillation.

**Pot still.** A traditional round-and-cylindrical still commonly used for making potable spirits.

**Racking.** The process of transferring the liquid portion of a wash or mash to a secondary fermentation vessel or still using a siphon. The siphon is also called the "racking cane."

**Raschig rings.** Tiny ceramic rings suspended in a still column to promote refluxing.

**Reflux distilling.** A distillation technique that allows vapors from boiling to rise, condense and trickle down into the mash or wash, where they are boiled again and refined as the still's temperature rises. Refluxing occurs inside a distillation column placed atop a still. As alcohol vapor rises through the column, it passes through copper or ceramic packing or copper plates that trigger condensation; that liquid falls back into the mash and is refluxed. Refluxing produces precise separation and cuts of the distillate and allows for very high temperature, high-proof distilling.

**Saccharification.** The result of adding the beta amylase enzyme to a mash that's undergone liquefaction. The process breaks small starch chains (dextrins) down to glucose, which is easily fermented by yeast.

**Second distillation.** The distillation of low wines into high-proof spirits.

**Spirit run.** *SEE* FINISHING RUN.

**Sour mash.** The addition of spent mash to new mash in order to trigger fermentation in the new batch and make flavors between each batch more consistent.

**Star San.** A popular no-rinse sanitizer for distillation equipment.

**Steeping.** Heating a fluid to increase flavor extraction from solids submerged in that fluid. In gin and absinthe making, spirits and low wines (respectively) are heated to macerate botanicals and extract flavors. In liqueur making, water and often spirits are heated to dissolve sugar and boost maceration of whole fruits or spices.

**Still.** A device designed to distill fermented low-alcohol solutions into high-proof spirits.

**Stripping run.** A rapid distillation designed to remove most of the water and some undesirable alcohols from a wash or mash. Precise cuts for heads, hearts and tails are not made in the stripping run.

**Sugar wash.** A simple wash made of sugar, water, yeast and yeast nutrients. This commonly is the base for simple moonshine.

**Tails.** Sometimes spelled "tales," this is the final portion of the distillate boiling off at about 203°F (95°C). Since tails do contain some ethanol, they typically are collected and reused in future distillation runs.

**Test cylinder.** A tall, narrow cylinder used to hold a sample of liquid for testing with a hydrometer or an alcoholmeter.

**Thumper/thump keg.** A sizable chamber (sometimes made from a wood barrel, sometimes metal) linked by copper tubing between a distiller and a condenser. It is used to increase the proof of distillate emerging from the still and eliminate the need for a finishing run. It gets its name from the thumping sound made by vapors coursing through condensed liquid inside it. It may also be referred to as a "doubler."

**Vodka.** A colorless, odorless and mildly flavored spirit that is distilled at 190°F (88°C) or higher.

**Wash.** A liquid mixture containing yeast and liquefied sugars (such as sugarcane, molasses or agave syrup) or granulated sugar dissolved in water for the purpose of creating alcohol through fermentation.

**Whiskey.** Any spirit distilled from a blend of grain and malted barley.

**White dog.** Unaged whiskey straight from the still.

**White lightning.** Unaged whiskey straight from an illegal still.

**Wort.** Liquid drained off after cooking malt to produce Scotch whiskey.

**Yeast.** A single-celled organism that, when added to a liquid solution, consumes sugar and converts it into alcohol. Countless strains of yeast exist, but only some are desirable for alcohol distillation. Each produces different flavors in distillates and each has a wide range of alcohol tolerance.

**Yeast nutrients.** Additives (usually potassium, phosphates and nitrogen) that ensure yeast propagation and maximum conversion of sugars to alcohol in a wash or mash.

# DISTILLING RESOURCES

## U.S.

**Adventures in Homebrewing**
www.homebrewing.org
Stills and parts, fermentation tools and equipment, ingredients, additives.

**American Still Co.**
www.americanstill.net
Stills, accessories, ingredients.

**Brewhaus**
www.brewhaus.com
817-750-2739
Stills and parts, accessories, additives, flavorings, ingredients, books, bottles, barrels.

**Clawhammer Supply**
www.clawhammersupply.com
828-419-0563
Still kits and parts, flavorings, recipes.

**Copper Moonshine Stills**
www.coppermoonshinestills.com
479-414-3220
Stills and parts, accessories, additives, flavorings and ingredients.

**Design2Brew**
www.design2brew.com
636-265-0751
Stills and parts, accessories, additives, flavorings and ingredients.

**Hillbilly Stills**
www.hillbillystills.com
270-334-3700
Stills, kits, fermentation tools, ingredients, infusion kits, distilling calculators.

**Homebrew Heaven**
www.store.homebrewheaven.com
425-355-8865
Small stills and parts, additives, ingredients.

**Mile High Distilling**
www.milehidistilling.com
303-987-3955
Stills and parts, fermentation tools, workshops, ingredients, essences, sanitizers, filters, additives, barrels.

**Moonshine Still Pro**
www.moonshinestillpro.com
Contact through website only.
Stills and parts.

**Moonshine Still USA**
www.moonshinestill.us
Contact through website only.
Stills, still kits and parts, fermentation tools, ingredients, additives, flavorings.

**Olympic Distillers**
www.olympicdistillers.com
Contact through website only.
Stills and parts, fermentation tools, ingredients, additives, flavorings, essences.

## CANADA

**Broken Oar Distilling Equipment**
www.brokenoardistillingequipment.com
250-713-6530
Stills and parts.

**Liquor Quik**
www.liquorquik.com
902-864-0100
Flavorings, additives, yeast, nutrients, sanitizers and more.

**Smiley's Home Distilling**
www.home-distilling.com
613-820-1069
Stills and parts, fermentation tools, ingredients, additives, flavorings, essences.

## AUSTRALIA

**Aussie Brewer**
www.aussiebrewer.com.au
+61 (0)7 40416608
Stills and parts, additives, essences, bases.

**Brewcraft**
www.liquorcraft.com.au
+61 (0)3 95791644
Stills and parts, additives, essences, ingredients.

**Distillery King**
www.distilleryking.com.au
Contact through website only.
Stills and parts, ingredients, essences.

**Hauraki BCL**
www.spiritsandbrewing.co.nz
+61 (0)7 492 64433
Stills and parts, accessories, ingredients, additives, essences.

**iBrew**
www.ibrew.com.au
+61 (0)7 55940388
Stills and parts, accessories, ingredients, additives, essences.

**iSpirits**
www.ispirits.com.au
+ 61 (0)3 9579 1644
Stills, filters, flavorings, additives.

**Malthouse**
www.malthouse.com.au
+61 (0)8 9361 6424
Stills and parts, accessories, ingredients, additives, essences.

## NEW ZEALAND

**Aqua Vitae**
www.aquavitae.co.nz
+64 (0)3 338 6224
Stills and parts, accessories, ingredients, additives, essences.

**Brewers Coop**
www.brewerscoop.co.nz
+64 (0)9 525 2448
Stills and parts, accessories, ingredients, additives, essences

**Hauraki BCL**
www.spiritsandbrewing.co.nz
North Island: +64 (0)9 8373311
South Island: +64 (0)3 688 0801
Stills and parts, accessories, ingredients, additives, essences.

**Spirits Unlimited**
www.spiritsunlimited.co.nz
+64 (0)3 688 0801
Stills and parts, accessories, ingredients, additives, essences.

**Still Spirits**
www.stillspirits.com
NZ: 0800 003548
AU: 1800 281 231
Small stills, flavors, filters, ingredients.

## UK

**Brewstore**
www.brewstore.co.uk
+44 (0)131 667 1296
Still kits, parts and tools.

**Homebrew Megastore**
www.homebrewmegastore.co.uk
+44 (0)1543 377 117
Stills and parts, accessories, ingredients.

**The Institute of Brewing & Distilling (IBD)**
www.ibd.org.uk
+44 (0)20 7499 8144

**Lovebrewing**
www.lovebrewing.co.uk
+44 (0)1246 279 382
Stills, still kits and parts, fermentation tools, ingredients, barrels.

# INDEX

## A
Abbott, Bud 18
absinthe 134, 140, 141, 150
airlock 22, 23
alcoholmeter 22, 23, 34
alpha amylase 36
Amaretto Disaronno 141
anaerobic environment 37, 38
Anti-Saloon League 11
Aperol 141
Applejack 120
Armagnac 120
Aviation gin 102

## B
bar tools 26–7
barley 107
barrel aging 82, 84, 88, 123
Bean, Heather 83, 88–9
Beefeater gin 102
beer 38
Bernstein, Barry 72–3
beta amylase 36
Big O Ginger Liqueur 140, 146–7
Bishop, Alan 121–2, 123, 128–9
bitters 140
Blake, Colin 102, 103
boiler 22, 33–4
Bombay Sapphire gin 102
botanicals
    gin 102–3, 106–7
    liqueurs 140
Boundary Oak Distillery 58–9
bourbon 10, 64
    barrels 82
    cocktail recipes 75, 76
bradwijn 120
brandy 8, 118–37
    aging 123
    American 120
    Applejack 120
    Armagnac 120
    barrel aging 123
    Calvados 120
    cocktail recipes 124, 130–7
    Cognac 120
    distillation 121–3
    distilled from wine 121, 127
    eau de vie 120
    fermentation 122
    filtering 123
    fine 120
    fresh fruit 121
    fruit juice 121
    grape varieties 122
    grappa 120
    hors d'age 121
    kirsch 120
    marc 120
    mash 118, 122
    Napoleon 121
    old 120
    pale 120
    pisco 120, 132
    schnapps 120, 140–1, 148–9
    special 120
    stripping run 123
    very special (VS) 120–1
    vintage 121
    VSOP 121
    wash 122
    XO 121
    yeast 122
Brettanomyces 40
brew pot 22
Brewhaus 16, 24
brix 65
Brown, Nate 114–15
Brown-Forman 11
Byron, Lord 80

## C
cachaça 97
calcium carbonate 41
Calvados 120
Campari 120
Canadian whiskey 64
cap 22
cap arm 22
carbon dioxide 37–8
Chambord 140
Chartreuse 140
cleaning equipment 39
climate 73
cocktail bar tools 26–7
cocktail recipes
    The Ampersand 130
    Angry Hound 53
    Blood and Sand 70
    Bolo 94
    Bourbon Rita 75
    Brandy Alexander 124
    Brandy and Ginger 124
    Cable Car 90
    Caipirinha 97
    Caprice 112
    Classic Daiquiri 86
    Commodore 94
    Corpse Reviver 133
    Cosmopolitan 54
    Crazy Tracy 117
    The Daisy 130
    Dark and Stormy 90
    French 75 117
    Group Hug Punch 61
    Hawaiian Sigh 48
    Hemingway Daiquiri 86
    Hot Buttered Rum 96
    Hurricane 93
    I'll Black Your Eye 79
    The Impressionist 137
    Irish Martini 48
    Lawn Dart 112
    Manhattan 69
    Martini 111
    Mint Julep 76
    Mojito 93
    Moscow Mule 57
    Negroni 104
    Old Fashioned 68
    Old Maid 108
    The Outsider 54
    Pisco Buck 132
    The Presbyterian 76
    The Prospector 57
    Ramos Gin Fizz 105
    Round Robin 134
    St. Charles Punch 137
    Salty Dog 53
    The Sazerac 70
    The Seelbach 75
    Tom Collins 108
    Vieux Carré 134
    Vieux Mot 111
    whiskey sour 79
    Wild Blue Yonder 61
cocktail shaker 27
Cognac 120
collector/collection vessel 22, 34
Colonial America 9
column still 20, 21, 22, 44, 59
commercial distilleries 14
condenser 20, 22, 34
    jacketed 22
condensing chamber 23
Copper & King's American Brandy 126–9, 140
corn whiskey 12
Corsair gin 102, 106–7
crème de cassis 140
Curacao 141
cuts, heads, hearts and tails 40, 47, 66–7, 84–5
Cynar 140

## D
Davis, Bryan 66
Death's Door gin 102
diacetyls 40
distillation
    brandy 121–3
    gin 102–3
    liqueurs 140
    rum 83–5
    vodka 44
    whiskey 66
doubler *see* thumper/thump keg

## E
eau de vie 120
enzymes 64
equipment

bar tools 26–7
cleaning 39
distilling 18–23
sanitizing 39
esters 40
ethanol 41
extractor 22

## F
Faulkner, William 6
fermentation 35–8, 59
    brandy 122
    exothermic, as 65
    rum 82–3
    stalled 41
    temperature 41
    under-fermented 37
    vodka 44
    whiskey 65
fermenter 22
Fields, W.C. 98
finishing run 20, 40
    rum 84–5
    vodka 45–6
    whiskey 66–7
Finland 11
Foster, Kathy, Bill and Chris 146–7
fruit 8, 118
    brandy 118, 120–3
    infusions 138–49
fusel oils 40

## G
Galliano 140
genever 8, 100–1
Gilby's gin 102
gin 8–9, 98–117
    botanicals 102–3, 106–7
    cocktail recipes 104–5, 108–12, 117
    distillation 102–3
    London dry 100, 101
    mash 100
    New Western 100, 102
    Old Tom 100, 101–2
    Plymouth 100, 102
    proofing down 107
    steeping and infusion 103
    vapour distillation 103, 107
glasses 26–7
glucoamylase enzymes 83

Goodin, Brent 58–9
Gordon's gin 102
graduated cylinder see test cylinder
grain 8, 9, 12, 62
    single-grain distillates 64
    whiskey 64
grain-neutral spirit 102, 107
grapes 107, 122, 127
grappa 15, 120

## H
Hamilton, Alexander 9
Haney, Mike 28–9, 33–4, 39, 46, 66, 85
hangovers 40
Harrison, Joel 150
Hauraki Brewing Co. 50–1
heat exchange condenser 34
Hendrick's gin 102
Hepola, Sarah 150
Hillbilly Stills 28–9
Hobby Distillers Association 12
Home Distillers Association (HDA) 24–5
hydrometer 22, 36–7, 38, 65

## I
infusions 103, 138–49
    potency 141, 142–3
Irish whiskey 64

## J
Jackson, Samuel 118
Jefferson, Thomas 10
jigger 26

## K
Kentucky Artisan Distillery 32
kettle 22
King, Justin 152–3
kirsch 120
Koji 38
krausen 37

## L
legality 14–17, 24–5
Libavias, Andreas 35
licensing 14
liqueur 140, 146–7
    vodka 44
louche 150
low wines 20, 66

lyne arm 22

## M
maceration 103, 107
malted barley 36
maltose 83
marc 120
Martin Miller's gin 102
mash 12, 36
    brandy 118, 122
    gin 102
    oxygenating 37, 41, 65
    smell 40
    temperature 37
    whiskey 64–5
mash pot 22
Maurer, David W. 10, 12
Mencken, H.L. 30
methanol 34
molasses 80, 82–3, 89
moonshining 10, 11–12, 14
Morris, Rick 16, 24–5, 38, 46, 84
Moscow Mule mug 27
muddler 27

## N
Napoleon 121

## O
O'Daniel, Brandon 122–3, 126–7
Old Forester Bourbon 11
Ole Smoky Moonshine 152–3
oxygen 37, 38, 41, 65

## P
packing 22
parrot 22–3, 46, 85
percolation 140
Pernod, Henri-Louis 150
pH 41
pisco 120, 132
poisonous distillate 34
polyphenols 40
pot still 20, 21, 23, 64, 152
    rum 83
    vodka 44
potassium carbonate 41
potatoes 44, 107
pressure 33
Prohibition 10, 11–12
proofing down 47, 85, 107, 142–3
Punt è Mes 140

## R
racking 38
racking cane 23, 38
raschig rings 23
reflux column 20, 22, 23, 59
reflux plates 23
Ridley, Nick 150
rum 8, 9, 35, 80–97
    barrel aging 82, 84, 88
    blending 85
    cocktail recipes 86, 90–7
    dark 82
    distillation 83–5
    fermentation 82–3
    finishing run 84–5
    golden 82
    infusions 140
    light 82
    molasses 80, 82–3, 89
    proofing down 85
    rhum agricole 82
    spiced 82, 140
    storing 85
    stripping run 83–4
    sugarcane 82
    wash 82–3
    white 82

## S
saccharification 36
saccharomyces bayanus 35
saccharomyces cerevisiae 35
safety 14, 16, 32–4, 59
sake 38
sanitizing equipment 39
schnapps 120, 140–1
    orange 148
    raspberry 149
Scotch whiskey 64
Sherman, Rob 29
shotgun condenser 34
sight glass 23
Smith, Clay 102, 106–7
sodium carbonate 41
solder 33
Solmonson, Lesley Jacobs 8–9, 100
speakeasy 11
specific gravity 41, 65
spirit run 20
Stalk & Barrel 72, 73
steeping 103, 140
still

assembly 33
column/reflux 20, 21, 22, 59
heat sources 33–4
leaks 33
output speed 41
pot 20, 21, 23
size 51
still cap 23
Still Waters Distillery 72–3
Stimson, Tripp 32–4, 40, 65, 67, 103
stirring spoon 23
stripping run 20, 40
    brandy 123
    rum 83–4
    vodka 44–5
    whiskey 66
sugar 36
    testing level of 36–7
    see also wash
sugar shine 12
sugarcane 82
Syntax Spirits Distillery 88–9

## T

taste, troubleshooting 40
taxation 9–10
tequila 140
test cylinder 23
Thomson Whisky 16
thumper/thump keg 23
triple sec 140
turbo yeast 36

## V

Van Gogh, Vincent 150
vapor cone 23
vapour, alcohol 32, 41, 45, 66
vapour basket 103, 107
Veach, Michael 8
vodka 8, 12, 14, 42–7
    blending 47
    cocktail recipes 48, 53, 61
    distillation 44
    fermentation 44
    filtering 47
    finishing run 45–6
    flavoring 51
    fruit-infused 142–4
    infusions 140
    liqueurs 44
    rum base spirit 83

storing 47
stripping run 44–5

## W

wash 12, 36, 59
    acidity 41
    brandy 122
    oxygenating 37
    specific gravity 41
    temperature 37, 41
Washington, George 9–10, 14
Wheeler, Peter 16, 36, 50–1
whiskey 8, 9–10, 59, 62–7
    aging 67
    barrels 67, 82
    blends 64
    Bourbon 10, 64
    Canadian 64
    cocktail recipes 68–70, 75–9
    color 67
    cooking 64
    corn 12
    distillation 66
    fermentation 65
    finishing run 66–7
    grain 62

    Irish 64
    mash 36, 64–5
    pot still 64
    Scotch 64
    single malts 64, 72–3
    storing 67
    stripping run 66
    wort 64–5
Whiskey Rebellion 9, 10
Wilde, Oscar 150
Willmott, Malcolm 50
worm 20, 23, 34
wort 64–5

## Y

yeast 35–6, 37–8, 59
    age 41
    brandy 122
    fall out 38
    pitching 36, 37

## Z

Zucca 140

# ACKNOWLEDGMENTS

There's no way to adequately thank the dozens of people who helped produce this book through the generous provision of their time: people like Tripp Stimson (Kentucky Artisan Distillery), Colin Blake (Distilled Spirits Epicenter), Mike Haney (Hillbilly Stills), Rick Morris (Brewhaus and the Hobby Distillers Association), and so many others who had better things to do than listen to a reporter's endless questions — yet they did. Even when I called back for clarification, none hesitated to explain and repeat themselves. To write that off as professional courtesy wouldn't be fair. These are just nice people — maybe it's the delicious liquor they distill that keeps them easygoing and cordial.

Thanks to the team at Quintet Publishing who organized the myriad details of this project and nudged me along with patience and grace. This experience reminds me why I don't want to self-publish anytime soon.

Thanks also to the authors who've researched this subject before me. The stack of books and research material I began with appeared a daunting read until I learned that what I had wasn't half of what's available on the history and how-to of hobby distilling. The hard part wasn't getting through the lot of it, the biggest challenge was saying, "Stop! Enough reading! Get writing!"

Thanks always and forever to my patient wife and son, who've seen me too little of late as I've disappeared into my office to write. Without God and the both of you in my life, such rigorous, vigorous effort wouldn't be worth it.

Steve Coomes
April 1, 2015